What does this painting have to do with math?

Swiss-born artist Paul Klee was interested in using color to express emotion. Here he created a grid, or array, of 35 colorful squares arranged in 5 rows and 7 columns. We will learn how an array helps us understand a larger shape by looking at the smaller shapes inside. Learning more about arrays will help us notice patterns and structure—an important skill for multiplication and division.

On the cover

Farbtafel "qu 1," 1930
Paul Klee, Swiss, 1879–1940
Pastel on paste paint on paper, mounted on cardboard
Kunstmuseum Basel, Basel, Switzerland

Paul Klee (1879–1940), *Farbtafel "qu 1"* (*Colour Table "Qu 1"*), 1930, 71. Pastel on coloured paste on paper on cardboard, 37.3 x 46.8 cm. Kunstmuseum Basel, Kupferstichkabinett, Schenkung der Klee-Gesellschaft, Bern. © 2020 Artists Rights Society (ARS), New York.

EUREKA MATH²™

GREAT MINDS

Great Minds® is the creator of *Eureka Math*®,
Wit & Wisdom®, *Alexandria Plan*™, and *PhD Science*®.

Published by Great Minds PBC.
greatminds.org

Printed in the USA

1 2 3 4 5 6 7 8 9 10   LSC   25 24 23 22 21

ISBN 978-1-64497-167-3

A Story of Units®

Units of Any Number ▸ 3

TEACH

Overview

Multiplication and Division with Units of 2, 3, 4, 5, and 10

Topic A

Conceptual Understanding of Multiplication

Students connect their understanding of equal groups and repeated addition to multiplication. They identify the number of groups, the number in each group, and the total within equal-groups models and arrays. They write multiplication equations to represent the equal groups and arrays. Students interpret the meaning of the factors as the number of groups and the number in each group and solve word problems.

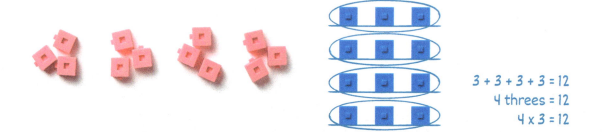

$$3 + 3 + 3 + 3 = 12$$
$$4 \text{ threes} = 12$$
$$4 \times 3 = 12$$

Topic B

Conceptual Understanding of Division

Students use equal-groups models and arrays to explore the two interpretations of division: measurement and partitive. They determine the total and either the number of groups or the number in each group based on the problem situation. They identify what is known and what is unknown, relate it to an unknown factor problem, and write a division equation. Students solve word problems involving division and make connections between multiplication and division.

$$\square \times 3 = 15$$

$$\underset{\text{total}}{15} \div \underset{\substack{\text{number in} \\ \text{each group}}}{3} = \underset{\substack{\text{number} \\ \text{of groups}}}{\square}$$

$$5 \times \square = 15$$

$$\underset{\text{total}}{15} \div \underset{\substack{\text{number} \\ \text{of groups}}}{5} = \underset{\substack{\text{number in} \\ \text{each group}}}{\square}$$

$$5 \times \square = 15$$

$$\underset{\text{total}}{15} \div \underset{\substack{\text{number} \\ \text{of rows}}}{5} = \underset{\substack{\text{number in} \\ \text{each row}}}{\square}$$

$$\square \times 3 = 15$$

$$\underset{\text{total}}{15} \div \underset{\substack{\text{number in} \\ \text{each row}}}{3} = \underset{\substack{\text{number} \\ \text{of rows}}}{\square}$$

Topic C

Properties of Multiplication

Students use the properties of multiplication to explore strategies they can use to multiply efficiently. They explore the commutative property of multiplication by skip-counting the rows and columns in arrays, which helps them build number sense while learning multiplication facts. Students use arrays and number bonds to model the distributive property when finding the products of unfamiliar facts.

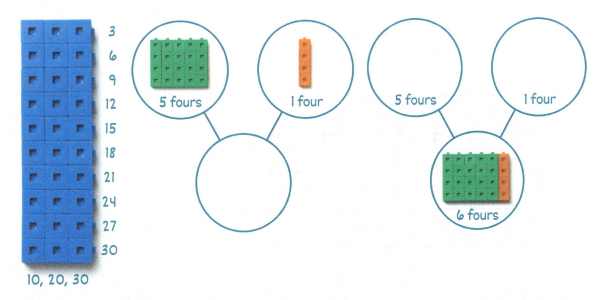

3
6
9
12
15
18
21
24
27
30

10, 20, 30

5 fours   1 four

5 fours   1 four

6 fours

After this Module:

Grade 3 Module 3

In grade 3 module 3, students apply conceptual understanding and use the commutative, distributive, and associative properties to extend their learning of multiplication and division to units of 6, 7, 8, 9, 0, 1, and two-digit multiples of 10. Students solve one- and two-step word problems involving the four operations.

Topic D

Two Interpretations of Division

Students solidify their understanding of the relationship between multiplication and division and express division as both unknown factor problems and division equations. Students describe the quotient as either the number of groups (as shown in the examples) or the size of each group and draw tape diagrams to represent the problems.

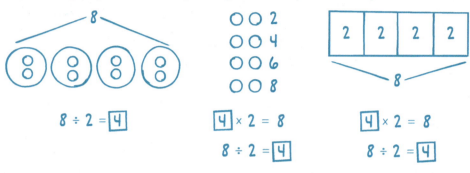

Topic E

Application of Multiplication and Division Concepts

Students apply the distributive property to complete multiplication and division problems and explore the foundations of the associative property of multiplication by breaking apart arrays into smaller arrays. Students solve two-step word problems using multiplication and division.

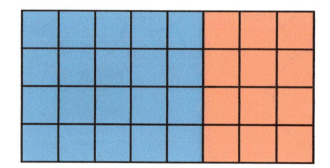

$$32 \div 4 = 5 + 3 = 8$$

20 12

Contents

Multiplication and Division with Units of $2, 3, 4, 5,$ and 10

Why

Multiplication and Division with Units of 2, 3, 4, 5, and 10

Why are multiplication and division concepts taught in module 1 and module 3?

Understanding and applying multiplication and division concepts are part of the major work of grade 3. Beginning the year with multiplication and division concepts and units of 2, 3, 4, 5, and 10 maximizes time for students to develop conceptual understanding and build fluency. In module 2, students have continued daily practice with multiplication and division through fluency activities. Students return to multiplication and division in module 3. They build upon the strong foundation established in module 1, now extending to more complex strategies and units of 6, 7, 8, 9, 0, and 1.

Multiplication concepts are the basis of much of the other work in grade 3 (e.g., area of plane figures, building fractions from unit fractions, and scaled bar and picture graphs). Beginning the year with multiplication allows for rich connections and enables multiplication to be the lens through which other concepts are explored.

How does the learning progress from module 1 to module 3?

Students' familiarity with skip-counting by twos, fives, and tens from earlier grade levels provides a natural starting point to establish the concepts of multiplication and division in module 1.

Early representations for multiplication and division in module 1 include equal groups and arrays, which are used to skip-count. Students learn the meaning of multiplication and division with familiar units of 5 and 10 and the smaller units of 2, 3, and 4, which allows for a smoother transition to more abstract representations, such as tape diagrams, before working with larger units.

Seeing the relationship between 10 and 5—that is, 10 as 5 doubled—provides a foundation for students' understanding of 4 as 2 doubled. In module 3, that understanding extends to 6 as 3 doubled and 8 as 4 doubled. The units of 6, 7, 8, and 9 are introduced in module 3 after students have developed some proficiency with smaller units and strategies based on the commutative and distributive properties. These strategies allow students to multiply and divide with larger units by creating easier problems using smaller and familiar units.

After students have developed some proficiency with the other single-digit factors, they can explore and understand the factors of 0 and 1 through patterns. Students learn why multiplication and division with 0 and 1 are unique and develop the meaning behind them to support understanding of the identity property in later grades.

Why do lessons focus on certain representations and tools for multiplication and division? Is it ok for students to use other representations and tools when they are not included in a lesson?

Most lessons include multiple representations and tools to facilitate access for all students. Allow students to demonstrate their understanding using representations and tools that make sense to them, even if those representations or tools are not a focus in the lesson.

Achievement Descriptors: Overview

Multiplication and Division with Units of $2, 3, 4, 5,$ and 10

Achievement Descriptors (ADs) are standards-aligned descriptions that detail what students should know and be able to do based on the instruction. ADs are written by using portions of various standards to form a clear, concise description of the work covered in each module.

Every module has its own set of ADs, and the number of ADs varies by module. Taken together, the sets of module-level ADs describe what students should accomplish by the end of the year.

ADs and their proficiency indicators support teachers with interpreting student work on

- informal classroom observations,
- data from other lesson-embedded formative assessment,
- Exit Tickets,
- Topic Quizzes, and
- Module Assessments.

This module contains the nine ADs listed.

3.Mod1.AD1

Represent a multiplication situation with a model and **convert** between several representations of multiplication.

Note: This excludes the creation of a multiplication situation from an expression, an equation, or a model, which is reserved for module 3.

3.OA.A.1

3.Mod1.AD2

Represent a division situation with a model and **convert** between several representations of division.

Note: This excludes the creation of a division situation from an expression, an equation, or a model, which is reserved for module 3.

3.OA.A.2

3.Mod1.AD3

Solve one-step word problems using multiplication and division within 100, involving factors and divisors 2–5 and 10.

Note: Only one factor needs to be 2–5 or 10.

3.OA.A.3

3.Mod1.AD4

Determine the unknown number in a multiplication or division equation involving factors and divisors 2–5 and 10.

Note: Only one factor needs to be 2–5 or 10.

3.OA.A.4

3.Mod1.AD5

Apply the commutative property of multiplication to multiply a factor of 2–5 or 10 by another factor.

3.OA.B.5

3.Mod1.AD6

Apply the distributive property to multiply a factor of 2–5 or 10 by another factor.

3.OA.B.5

3.Mod1.AD7

Represent and **explain** division as an unknown factor problem.

3.OA.B.6

3.Mod1.AD8

Multiply and **divide** within 100 fluently with factors 2–5 and 10, recalling from memory all products of two one-digit numbers.

Note: Only one factor needs to be 2–5 or 10.

3.OA.C.7

3.Mod1.AD9

Solve two-step word problems.

Note: For module 1, in multiplication or division problem types, at least one factor or the divisor must be 2–5 or 10.

3.OA.D.8

The first page of each lesson identifies the ADs aligned with that lesson. Each AD may have up to three indicators, each aligned to a proficiency category (i.e., Partially Proficient, Proficient, Highly Proficient). While every AD has an indicator to describe Proficient performance, only select ADs have an indicator for Partially Proficient and/or Highly Proficient performance.

An example of one of these ADs, along with its proficiency indicators, is shown here for reference. The complete set of this module's ADs with proficiency indicators can be found in the Achievement Descriptors: Proficiency Indicators resource.

ADs have the following parts:

- **AD Code:** The code indicates the grade level and the module number and then lists the ADs in no particular order. For example, the first AD for grade 3 module 1 is coded 3.Mod1.AD1.

- **AD Language:** The language is crafted from standards and concisely describes what will be assessed.

- **AD Indicator:** The indicators describe the precise expectations of the AD for the given proficiency category.

- **Related Standard:** This identifies the standard or parts of standards from the Common Core State Standards that the AD addresses.

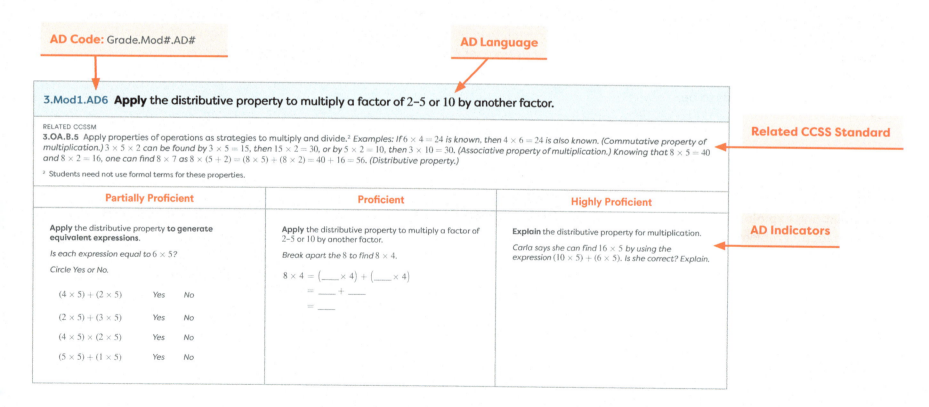

AD Code: Grade.Mod#.AD#

AD Language

Related CCSS Standard

AD Indicators

3.Mod1.AD6 **Apply** the distributive property to multiply a factor of 2–5 or 10 by another factor.

RELATED CCSSM

3.OA.B.5 Apply properties of operations as strategies to multiply and divide.² *Examples: If* $6 \times 4 = 24$ *is known, then* $4 \times 6 = 24$ *is also known. (Commutative property of multiplication.)* $3 \times 5 \times 2$ *can be found by* $3 \times 5 = 15$, *then* $15 \times 2 = 30$, *or by* $5 \times 2 = 10$, *then* $3 \times 10 = 30$. *(Associative property of multiplication.) Knowing that* $8 \times 5 = 40$ *and* $8 \times 2 = 16$, *one can find* 8×7 *as* $8 \times (5 + 2) = (8 \times 5) + (8 \times 2) = 40 + 16 = 56$. *(Distributive property.)*

² Students need not use formal terms for these properties.

Partially Proficient	Proficient	Highly Proficient
Apply the distributive property **to generate equivalent expressions**. *Is each expression equal to* 6×5? *Circle Yes or No.* $(4 \times 5) + (2 \times 5)$ Yes No $(2 \times 5) + (3 \times 5)$ Yes No $(4 \times 5) \times (2 \times 5)$ Yes No $(5 \times 5) + (1 \times 5)$ Yes No	**Apply** the distributive property to multiply a factor of 2–5 or 10 by another factor. *Break apart the 8 to find* 8×4. $8 \times 4 = (\underline{} \times 4) + (\underline{} \times 4)$ $= \underline{} + \underline{}$ $= \underline{}$	**Explain** the distributive property for multiplication. *Carla says she can find* 16×5 *by using the expression* $(10 \times 5) + (6 \times 5)$. *Is she correct? Explain.*

Topic A
Conceptual Understanding of Multiplication

Topic A lessons provide students with time and space to build conceptual understanding of multiplication, through concrete and pictorial exploration, with equal groups and arrays. Students expand their foundational understandings of multiplication from grade 2. The relationship between the number of groups, the number in each group, and the total is established, and students repeatedly practice identifying the number of groups and the number in each group. Building conceptual understanding in this topic helps establish tools and strategies that students can use to make sense of multiplication facts and to recognize situations where they can multiply to solve problems.

The topic opens with students counting a collection of objects. This serves as an informal formative assessment of the foundational understanding of multiplication from grade 2 and provides teachers with insight as to how students organize, count, and represent the collections. In addition to its mathematical content goals, the lesson's format sets the stage for students to work with and learn from each other throughout the year.

Unit form and skip-counting provide bridges from repeated addition to multiplicative thinking. Equal groups representations progress from concrete objects, to drawings, to arrays, and finally to tape diagrams. Regardless of representation, the number of groups and number in each group are consistently related to unit form and multiplication expressions and equations. Precise use of language and symbols are emphasized throughout the topic to maintain consistency between similar and interchangeable terminology (e.g., the convention of multiplication as the number of groups times the number in each group and the use of terms such as *factor, product, multiply, multiplication,* and *times*).

In topic B, students build conceptual understanding of division by using what they know about multiplication and the relationship of the number of groups, the size of the groups, and the total.

Progression of Lessons

Lesson 1

Organize, count, and represent a collection of objects.

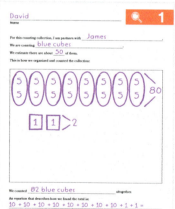

If I organize my collection into equal groups, I can skip-count and then add the extras to find the total.

Lesson 2

Interpret equal groups as multiplication.

3 + 3 + 3 + 3 = 12
4 threes = 12
4 × 3 = 12

Multiplication is another way to represent repeated addition. *I use unit form and write multiplication equations to represent equal groups. The symbol × is used to show multiplication and to write multiplication expressions and equations.*

Lesson 3

Relate multiplication to the array model.

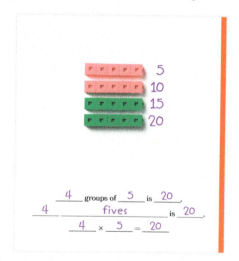

  4 groups of 5 is 20 .
  4   fives   is 20 .
    4 × 5 = 20

I organize equal groups in an array or tape diagram. Then I can see the number of groups and the number in each group and I can write a multiplication equation.

Lesson 4

Interpret the meaning of factors as number of groups or number in each group.

The numbers I multiply in a multiplication problem are called factors. They describe *the number of groups* and *the number in each group*. In the tape diagram, I see 7 groups with 5 in each group. The tape diagram represents $7 \times 5 = 35$.

Lesson 5

Represent and solve multiplication word problems by using drawings and equations.

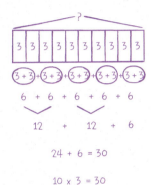

$24 + 6 = 30$

$10 \times 3 = 30$

There are 30 people on the roller coaster.

Using Read–Draw–Write helps me make sense of problems. When I see the drawings and strategies my classmates used to solve the problem, I learn more strategies to use next time.

1

Organize, count, and represent a collection of objects.

Name _____

 1

1. What unit did you use to count your collection? Explain why you chose that unit.

   Sample:

   I chose fives. Counting by fives is quick. I found the total using fewer counts than I would have if I had counted by ones.

2. If you counted your collection again, would you choose the same unit? Explain.

   Sample:

   I would use a different unit. I would use tens because I could find the total using fewer counts than I used when I counted by fives.

7

Lesson at a Glance

This student-driven lesson provides and opportunity to gather formative assessment data as students work with counting collections. Students decide how to organize, count, and represent the items. They analyze the work of others and discuss efficient strategies as a class.

There is no Problem Set for this lesson. Instead, use classroom observations and the classwork to analyze student thinking after the lesson. The Exit Ticket for this lesson serves as an opportunity for students to reflect on their counting strategies.

Key Question

- How can we use groups to help us organize?

Achievement Descriptor

This lesson is foundational to the work of grade 3 and builds from 2.NBT.A.2. Its content is intended to serve as a formative assessment and is therefore not included on summative assessments in grade 3.

Agenda

Fluency 5 min

Launch 10 min

Learn 35 min

- Organize, Count, and Record
- Share, Compare, and Connect

Land 10 min

Materials

Teacher

- 100-bead rekenrek
- Color tiles, plastic, 1 inch (60–150)
- Interlocking cubes, 1 cm (100)
- Computer or device*
- Projection device*
- *Teach* book*

Students

- Color tiles, plastic, 1 inch (60–150 per student pair)
- Interlocking cubes, 1 cm (60–150 per student pair)
- Organizational tools
- Dry-erase marker*
- Eraser*
- *Learn* book*
- Pencil*
- Personal whiteboard*
- Personal whiteboard eraser*

These materials are only listed in lesson 1. Ready these materials for every lesson in this module.

Lesson Preparation

- Create collections of interlocking cubes or color tiles that have between 60 and 150 objects (per student pair). Organize each collection in a bag or small box. Although this lesson uses interlocking cubes and color tiles, you can incorporate other items of high interest.

- Display tools for students to choose from to help organize their counts. Tools may include envelopes, cups, bags, rubber bands, or graph paper.

Fluency

Counting on the Rekenrek by Tens

Materials—T: Rekenrek

Students count by tens in unit and standard form to develop an understanding of multiplication.

Show students the rekenrek. Start with all the beads to the right side.

Say how many beads there are as I slide them over.

Slide the top row of beads all at once to the left side.

10

The unit is 10. In unit form, we say 1 ten. Say 10 in unit form.

1 ten

Slide the second row of beads all at once to the left side.

How many beads are there now? Say it in unit form.

2 tens

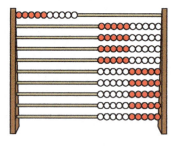

Student View

Continue sliding over each row of beads all at once as students count.

3 tens, 4 tens, 5 tens, 6 tens, 7 tens, 8 tens, 9 tens, 10 tens

Slide all the beads back to the right side.

Now let's practice counting by tens in standard form. Say how many beads there are as I slide them over.

Let's start at 0. Ready?

Slide over each row of beads all at once as students count.

0, 10, 20, 30, 40, 50, 60, 70, 80, 90, 100

Counting the Math Way by Tens

Students relate counting on the rekenrek to counting the math way to develop a strategy for multiplying beginning in lesson 2.

Let's count the math way. Each finger represents 10, just like a row on the rekenrek.

Face the students and direct them to mirror you. Show a fist with your right hand, palm facing out.

Show me your left hand. Make a fist like me. That's 0.

Now, raise your right pinkie.

Show me your left pinkie. That's 10.

Differentiation: Support

Students with fine motor delays may find it easier to use their fingers when they lay their hands flat on the desk or floor. The flat surface helps them hold out the fingers they want up and keep the others tucked in.

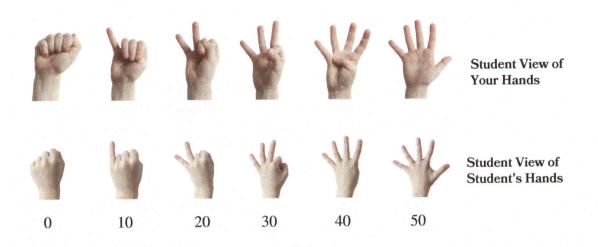

Student View of Your Hands

Student View of Student's Hands

| 0 | 10 | 20 | 30 | 40 | 50 |

Let's put up the very next finger.

Raise your right ring finger. Students raise their left ring finger.

That's 20.

Put up the next finger. 30.

Now that students understand the routine, switch to having them say the count as they show fingers. Guide students to continue counting the math way by tens to 100, then back down to 0.

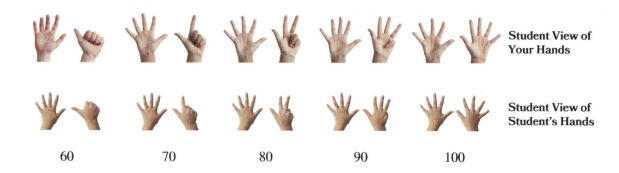

Student View of Your Hands

Student View of Student's Hands

60   70   80   90   100

Launch ⏱ 10

Materials—T: Counting collection; S: Counting collection, organizational tools

Students estimate the total of a class collection and prepare to count one of their own.

Gather students and display a counting collection.

Language Support

Consider using strategic, flexible grouping throughout the module based on students' mathematical and English language proficiency. Grouping suggestions follow:

- Pair students who have different levels of mathematical proficiency.

- Pair students who have different levels of English language proficiency.

- Join two pairs to form small groups of four.

As applicable, complement any of these groupings by pairing students who speak the same native language.

Invite open-ended discussion by asking students what they notice or wonder about the collection.

Then ask students to estimate the total number of objects in the collection. Ask questions such as:

- What prediction, or estimate, would be too big? Why?
- What prediction, or estimate, would be too small? Why?

Briefly orient them to the materials and procedure for the counting collection activity:

- Partners will collaborate to count a collection.
- Partners will make their own recordings to show how they counted.
- Partners may use organizational tools. Organizational tools may include readily available classroom items such as cups, rubber bands, personal whiteboards, etc.

Partner students and distribute a different counting collection to each pair.

Before they begin to count, invite partners to work together to predict how many objects are in their collection. Have them write down their estimates. Then encourage them to talk about how they will organize their collections to count.

Invite students to select organizational tools they would like to use, with the understanding that tools may be exchanged as plans are refined.

Transition to the next segment by framing the work.

Today, we will count our collections and record the ways we organize and count.

Teacher Note

Plan for what students should do when they finish counting their collection and recording how they counted:

- Try another way to organize and count.
- Switch collections with another student pair, and count to confirm the total.
- Explain their recording to another pair.
- Clean up and get another collection.

Learn 35

Organize, Count, and Record

Materials—S: Counting collection, organizational tools

Partners organize and count a collection and record their process.

Ask partners to begin counting their collections. Circulate and notice how students engage in the following behaviors:

Organizing: Strategies may include counting a scattered configuration, separating counted and uncounted objects, lining up objects as they are counted, making equal groups, creating 5-groups, forming arrays, and writing expressions or equations. Students may also organize their collections using attributes that do not support counting efficiently, such as by color or size.

Counting: Students may count by ones, twos, fives, or tens. Some may count subgroups and then add to find the total.

Recording: Recordings may include drawings, numbers, expressions, equations, and written explanations.

Circulate and use questions and prompts such as the following to assess and advance student thinking:

- Show and tell me what you did.
- How can you organize your collection to make it easier to count?
- How does the way you organized your collection make it easier to count?
- How did you keep track of what you already counted and what you still needed to count?
- How close was your estimate to your actual count?

Promoting the Standards for Mathematical Practice

Students look for and make use of structure (MP7) as they decide how to organize their counting collections to make them easier to count.

Ask the following questions to promote MP7:

- What's another way to organize your collection that would help you count?
- How does what you know about counting by tens help you to count your collection?

Differentiation: Support

If students sort by size, color, or other ways not related to equal groups, help them transition to more efficient ways of organizing and counting. Use prompts such as the following:

- How can you organize your collection to help you count?
- What organizational tools can help you count?
- Let's visit another group to see what helps them count their collection.

Select two or three pairs of students to share their counting work in the next segment. If possible, take pictures to show the class in the next segment. The samples show possible strategies. They demonstrate:

- making equal groups and counting on by that unit (e.g., grouping by fives and skip-counting),

- counting by a basic unit (e.g., sorting by an attribute and then adding the totals of the groups), and

- organizing groups as an array and counting on by that unit (e.g., making an array with ten in each row and knowing the total without having to skip-count).

Teacher Note

Students can progress at different paces with counting and recording, so they may count and record with different levels of sophistication.

Group and Count by Fives

Group by Color and Count by Ones

Organize in an Array and Recognize the Total

For this counting collection, I am partners with _____.

We are counting _____.

We estimate there are about _____ of them.

This is how we organized and counted the collection:

We counted _____ altogether.

An equation that describes how we found the total is:

_____

Action & Expression

Write one thing that worked well for you and your partner. Explain why it worked well.

It worked well for me and my partner to connect the cubes into groups of 10 to organize our collection. We both agreed that 10 was a number we could count by.

Write one challenge you had. How did you work through the challenge?

At first we thought we should make groups of 20, but the sticks kept breaking, and we lost track of our count. We decided to make sticks of 10 instead.

Share, Compare, and Connect

Students discuss strategies for organizing and compare the efficiency of each.

Gather the class to view the selected work samples and lead a discussion. Invite the selected pairs to share their counting process. The following dialogue models a sample discussion.

Teacher Note

Consider reserving time for the class to engage in discussion after partners have time to complete the self-reflection questions. Development of metacognitive strategies may support students in understanding how they learn best and help them self-monitor their progress.

Group and Count by Fives (David and James's Way)

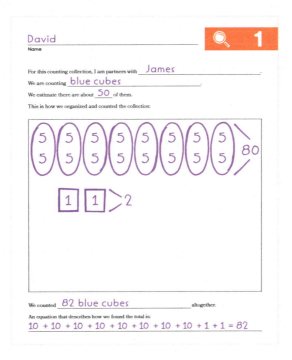

David
Name

For this counting collection, I am partners with __James__.

We are counting __blue cubes__.

We estimate there are about __50__ of them.

This is how we organized and counted the collection:

5 5 5 5 5 5 5 5
5 5 5 5 5 5 5 5 ⟩ 80

1 1 ⟩ 2

We counted __82 blue cubes__ altogether.

An equation that describes how we found the total is:
10 + 10 + 10 + 10 + 10 + 10 + 10 + 10 + 1 + 1 = 82

UDL: Representation

Consider creating a three-column chart and, as partners share, recording each strategy. After all pairs have shared, compare the organization of the count, the method for finding the total, and the efficiency of each strategy. For example, emphasize that each pair organized in a different way (by color, by 5-groups, and in rows of 10) and found the total in different ways (counted on, skip-counted, and used a known fact).

This sample uses a skip-count.

Invite partners to share.

How did you know the total?

We counted by fives.

Why did you decide to count that way?

We know how to count by fives and tens, so we made groups of 5 with our cubes and put 2 fives together to make tens on the sheet.

Can you count your collection a different way? How?

Yes, we can put 2 tens together and skip-count by twenties. Then we can add what's left.

Ask students to raise their hand if they used a skip-count.

Group by Color and Count by Ones (Ivan and Jayla's Way)

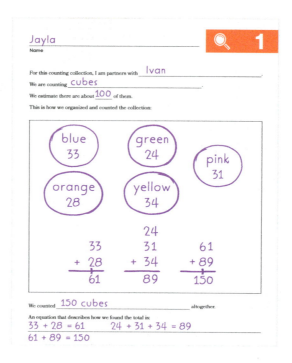

Jayla
Name

For this counting collection, I am partners with **Ivan**

We are counting **cubes**

We estimate there are about **100** of them.

This is how we organized and counted the collection:

blue
33

green
24

pink
31

orange
28

yellow
34

$$24$$
$$33 \qquad 31 \qquad 61$$
$$+\ 28 \qquad +\ 34 \qquad +\ 89$$
$$\overline{61} \qquad \overline{89} \qquad \overline{150}$$

We counted **150 cubes** altogether.

An equation that describes how we found the total is:

33 + 28 = 61   24 + 31 + 34 = 89

61 + 89 = 150

Teacher Note

As students count, they exhibit different levels of sophistication in their counting strategies. By selecting students to share their work, students with less sophisticated counting strategies have an opportunity to hear new ideas. If time allows, encourage students to count their collection a second time using a strategy they heard from another group.

Note that sorting by color is a common strategy but that it is not always efficient.

Invite partners to share their counting strategy.

Could we use David and James's way, a skip-count, to find the total of this collection?

Yes. We could count each color group by fives or tens.

How would the recording change if we skip-counted by fives or by tens?

We wouldn't have to add all the numbers together. We could just count in our head.

Organize in an Array and Recognize the Total (Mia and Amy's Way)

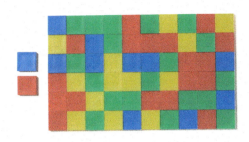

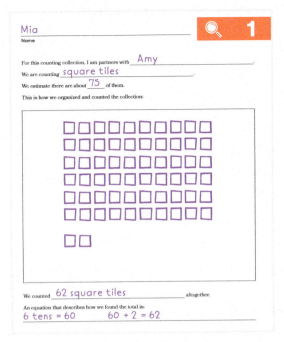

Mia
Name

🔍 **1**

For this counting collection, I am partners with **Amy**

We are counting **square tiles**

We estimate there are about **75** of them.

This is how we organized and counted the collection:

We counted **62 square tiles** altogether.

An equation that describes how we found the total is:

6 tens = 60   60 + 2 = 62

Invite partners to share.

How did your organization help you count?

We made 6 rows of 10 tiles. We know that 6 rows of 10 is 60, so we didn't have to count all the tiles or skip-count. We had 2 extra tiles, so we added 2 to 60 and got 62.

Invite students to turn and talk about how Mia and Amy's strategy could be used for 8 rows of 10.

Some students might make the connection that they can find the total of their array by multiplying, but this is not an expectation of students at this point in the year. If students make the connection, invite them to share their thinking. Formal introduction of multiplication using different units begins in lesson 2.

If time permits, have students show a new partner their recording and explain their work.

Use the following prompts to guide a discussion about how the organization of a collection helps in finding the total.

What were you successful with when counting?

I organized my cubes into equal groups so I could skip-count to find the total.

What did you find challenging about counting?

At first, I counted by twos. It was hard to keep track of my count. Then I switched to making and counting groups of 5 and that was much easier.

Did you see something you'd like to try next time we count collections? Why do you want to try it?

Next time, I'll use groups of 5 again, but I'll organize the groups better so that I know which ones I've already counted and which ones I haven't.

I'd like to use cups to put my collection in the next time. We counted cubes and they were hard to line up. We could just count 10 and put 10 in each cup. Then we could skip-count by tens to find the total.

Next time, I'd like to try multiplication to find the total number.

How does organizing help you count?

It helps me to keep track of which things I've counted, and it makes it faster to count. If I make a mistake, I don't have to start all over again.

Land 🔟

Debrief 5 min

Objective: Organize, count, and represent a collection of objects.

Display *Flower Vendor*, 1949, by Diego Rivera.

This painting is called *Flower Vendor*. The artist who painted this is named Diego Rivera. It is one of many paintings he made of calla lily flowers.

Use the following questions to help students engage with the art:

- What do you notice in the painting?
- What do you wonder?

Guide students to think about the painting in terms of their experience with the counting collection. Tell the class that the children in the painting are making bundles of flowers for the woman to carry. Then ask:

Do you think they are making equal groups? Why or why not?

No, they're not making equal groups. It's easy to tell just by looking that the bundles are different sizes.

Why might it be useful to make equal groups of flowers?

It might be easier for the woman to carry equal groups. If she had one giant bundle and one small bundle, it could be hard to manage.

To find the total

To count them

Why might the woman need to know how many flowers are in the collection?

Maybe she's planning to sell them. The name of the painting is *Flower Vendor*.

Maybe she'll give them away, and she wants to be fair to her friends.

Teacher Note

The focal point of this painting is the woman in the center. Rivera draws the eye to her by using bright colors in her dress, contrasting her head against the white flowers, and using the accents on her dress to frame her face.

Notice that on the left side of the painting, the girl's clothing mirrors the linearity of the stems. On the right side of the painting, the girl's hair and clothing mirror the heart-like shape of the flower petals.

In the center of the painting, eyes may be drawn to the parallel lines that are formed by the angle of the boy's back and the stems of the two flower bundles.

How might knowing the number of flowers in each group help?

She could add them all up.

If they are equal groups, she can skip-count.

As time allows, use the following questions to deepen students' exploration of the art:

- What catches your eye in this painting? Why do you think you focused on that?
- Notice how the stems of the calla lily flowers are long and straight without leaves. Where else in the painting do you see lines like that?

Exit Ticket 5 min

Provide up to 5 minutes for students to complete the Exit Ticket. It is possible to gather formative data even if some students do not complete every problem.

Interpret equal groups as multiplication.

Name _____

✉ **2**

Use the equal groups for parts (a)–(d).

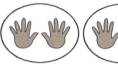

a. How many groups of 10 fingers are there?

___4___ groups

b. Fill in the blanks to show the total number of fingers.

___10___ + ___10___ + ___10___ + ___10___ = ___40___

c. How many tens are there?

___4___ tens

d. Fill in the blanks to match the picture.

___4___ × 10 = ___40___

13

Lesson at a Glance

Students find the total of equal groups using skip-counting and repeated addition. Students describe the groups in unit form and write related multiplication equations. This lesson formalizes the terms *multiply* and *multiplication* and introduces the multiplication symbol, ×.

Key Questions

- How is writing a multiplication equation more efficient than writing a repeated addition equation?
- What is the relationship between equal groups, repeated addition, unit form, and multiplication equations?

Achievement Descriptor

3.Mod1.AD1 **Represent** a multiplication situation with a model and **convert** between several representations of multiplication. (3.OA.A.1)

Agenda

Fluency 10 min

Launch 10 min

Learn 30 min

- Relate Equal Groups to Multiplication
- Represent Equal Groups with Multiplication Equations
- Problem Set

Land 10 min

Materials

Teacher

- 100-bead rekenrek

Students

- Interlocking cubes, 1 cm (12)

Lesson Preparation

Prepare 12 interlocking cubes of one color for each student.

Fluency ⑩

Ready, Set, Add

Students find the total and say an addition equation to maintain fluency within 10 from grade 1.

Let's play Ready, Set, Add.

Have students form pairs and stand facing each other.

Model the action: Make a fist and shake it on each word as you say, "Ready, set, add." At "add," open your fist, and hold up any number of fingers.

Tell students that they will make the same motion. At "add" they will show their partner any number of fingers. Consider doing a practice round with students.

Clarify the following directions:

- To show zero, show a closed fist at "add."

- Try to use different numbers each time to surprise your partner.

Each time partners show fingers, have them both say the total number of fingers. Then have each student say the addition equation, starting with the number of fingers on their own hand. See the sample dialogue under the photograph.

Circulate as students play the game to ensure that each student is trying a variety of numbers.

Partners A and B: "6"

Partner A: "$4 + 2 = 6$"

Partner B: "$2 + 4 = 6$"

Differentiation: Challenge

Students who demonstrate fluency adding within 10 can be challenged to add within 20. Encourage each partner to use both of their hands to show a number of fingers.

Counting on the Rekenrek by Tens

Materials—T: Rekenrek

Students count by tens in unit and standard form to develop an understanding of multiplication.

Show students the rekenrek. Start with all the beads to the right side.

Say how many beads there are as I slide them over.

Slide the top row of beads all at once to the left side.

10

The unit is 10. In unit form, we say 1 ten. Say 10 in unit form.

1 ten

Slide the second row of beads all at once to the left side.

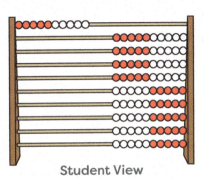

Student View

How many beads are there now? Say it in unit form.

2 tens

Continue sliding over each row of beads all at once as students count.

3 tens, 4 tens, 5 tens, 6 tens, 7 tens, 8 tens, 9 tens, 10 tens

Slide all the beads back to the right side.

Now let's practice counting by tens in standard form. Say how many beads there are as I slide them over.

Let's start at 0. Ready?

Slide over each row of beads all at once as students count.

0, 10, 20, 30, 40, 50, 60, 70, 80, 90, 100

Teacher Note

Take care not to count or "mouth" the words along with students. Students may learn to mimic the teacher rather than focus on number order.

Counting the Math Way by Tens

Students construct a number line with their fingers while counting aloud to develop a strategy for multiplying.

Let's count the math way. Each finger represents 10, just like a row on the rekenrek.

Face the students and direct them to mirror you. Show a fist with your right hand, palm facing out.

Show me your left hand. Make a fist like me. That's 0.

Now, raise your right pinkie.

Show me your left pinkie. That's 10.

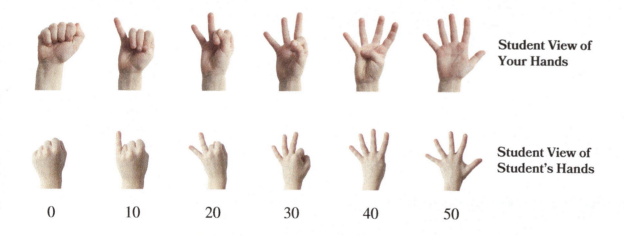

Student View of
Your Hands

Student View of
Student's Hands

| 0 | 10 | 20 | 30 | 40 | 50 |

Let's put up the very next finger.

Raise your right ring finger. Students raise their left ring finger.

That's 20.

Put up the next finger. 30.

Now that students understand the routine, switch to having them say the count as they show fingers. Guide students to continue counting the math way by tens to 100, then back down to 0.

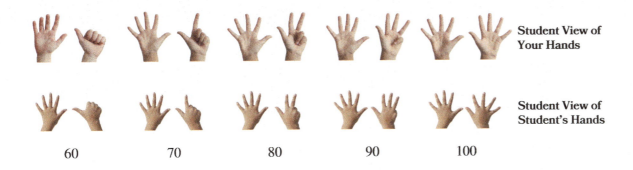

Student View of Your Hands

Student View of Student's Hands

60   70   80   90   100

 **Launch** 10

Students determine an efficient way to organize and count an unknown number of objects.

Gather the class and invite 10 students to stand in front of the room. Ask the class:

How many students are standing?

How many arms does each student have?

How many groups of 2 arms are there?

What addition expression can we write to represent our groups of 2?

$2 + 2 + 2 + 2 + 2 + 2 + 2 + 2 + 2 + 2$

Teacher Note

If counting student arms could be uncomfortable for your class, consider using eyes, ears, or shoes instead.

Write the addition expression.

How many twos do we have?

10 twos

Write 10 twos.

Ask students to work with a partner and add to find the total.

What is the value of 10 groups of 2?

20

Write = 20. Write 10 groups of 2 is 20.

Instead of adding, what is another way we can find the total of 10 twos?

Skip-count by twos.

Lead the class in skip-counting to find the total number of arms the students have.

Which way is more efficient: repeated addition of 2 until we have 10 twos or skip-counting by twos 10 times?

Ask students to return to their seats.

Invite students to turn and talk about how each of the three statements relates to the skip-count.

Transition to the next segment by framing the work.

Today, we will find a more efficient way to represent repeated addition.

$$2 + 2 + 2 + 2 + 2 + 2 + 2 + 2 + 2 + 2 = 20$$

10 twos

10 groups of two is 20.

> **Teacher Note**
>
> Throughout the lesson, equal group representations are used with familiar terms from grade 2, such as repeated addition, unit form, and skip-counting, to support students in conceptualizing multiplication.

Learn

Relate Equal Groups to Multiplication

Materials—S: Cubes

Students represent equal groups with repeated addition, unit form, and a multiplication equation.

Distribute 12 cubes to each student and ask them to take out their whiteboards.

Ask students to use their cubes to make equal groups of 3. Provide students time to work.

How many equal groups of 3 did you make?

You made 4 equal groups of how many cubes?

4 equal groups of 3 cubes is how many cubes altogether?

Ask students to write a repeated addition equation on their whiteboards to represent the groups.

What repeated addition equation represents the equal groups?

Write the repeated addition equation $3 + 3 + 3 + 3 = 12$.

How many threes did we add to make 12?

Write the unit form, 4 threes = 12, below the addition equation.

$3 + 3 + 3 + 3 = 12$
$4 \text{ threes} = 12$
$4 \times 3 = 12$
Multiply: 4 times 3 equals 12

How many times do you see a group of three?

Write $4 \times 3 = 12$.

We see a group of three 4 times. 4×3 is another way to write $3 + 3 + 3 + 3$ or 4 threes. These statements all represent the equal groups.

Invite students to think–pair–share about how $4 \times 3 = 12$ relates to repeated addition and unit form.

They are all groups of 3 and the answer is 12.

You add three 4 times because there are 4 groups of 3. That's how you get 4×3.

$4 \times 3 = 12$ is a shorter way to write the long addition equation.

When we have equal groups, *multiplication* is another way to represent repeated addition. Instead of repeatedly adding the same number, we *multiply* the number in each group by the number of groups. The times symbol is used to show multiplication.

Below the multiplication equation, write multiply: 4 times 3 equals 12.

Repeat the sequence to make equal groups of 2, 4, and 6 using 12 cubes. Continue to reinforce the relationship between repeated addition, unit form, and multiplication equations.

Represent Equal Groups with Multiplication Equations

Students represent pictorial equal groups with repeated addition and multiplication equations.

Show the picture of groups of 2 cubes. Invite students to turn and talk about how they know the groups are equal.

How many times do you see groups of 2?

I see two 5 times.

Ask students to work with a partner to write a related repeated addition equation and a multiplication equation for the picture.

Invite students to turn and talk about how the addition and multiplication equation represents the equal groups.

Show the picture of groups of cubes and invite students to think–pair–share about if they agree or disagree that the equation accurately represents the groups.

I disagree because my addition equation equals 13, not 15.

The last group doesn't have 5 cubes, so you can't say there are 3 fives.

You can multiply when the groups are equal. These groups aren't equal, so the picture doesn't show 3×5.

3 x 5 = 15

I hear most of you disagreeing because the groups are not equal. To multiply, you must have equal groups.

What should I change in the picture to make it match the multiplication equation?

You should put 2 more cubes in the group that has 3 cubes.

UDL: Representation

Consider providing additional examples and nonexamples to emphasize that to multiply you must have equal groups. For example, display three groups of cubes that are not equal: the first group with 5 cubes, the second group with 5 cubes, and the third group with 4 cubes. Then have students write an addition expression to represent the total. Students should write $5 + 5 + 4$. Discuss why $5 + 5 + 4$ cannot be represented with multiplication.

Repeat the process with three equal-sized groups and discuss why these groups can be represented with multiplication.

Alternatively, provide a repeated addition equation and ask students to represent it with manipulatives or ask students to generate the examples themselves.

Problem Set

Differentiate the set by selecting problems for students to finish independently within the timeframe. Problems are organized from simple to complex.

Land

Debrief 5 min

Objective: Interpret equal groups as multiplication.

Initiate a class discussion using the prompts below. Encourage students to restate their classmates' responses in their own words.

How does thinking about how many times we see equal groups help us write a multiplication equation?

It helps me think about unit form and that helps me to write a multiplication equation. So, if I see a group of two 7 times, that's 7 twos which is the same as 7×2.

Why is writing a multiplication equation more efficient than writing a repeated addition equation?

When I write a repeated addition equation, I write the same number over and over again. When I write a multiplication equation, I think about the number of groups and the number in each group and then write an equation using only those two numbers.

Is it always more efficient to write a multiplication equation?

Sometimes it's not more efficient, like when I have 2 groups of 3. Writing 2×3 isn't any faster than writing $3 + 3$.

What is the relationship between equal groups, repeated addition, unit form, and multiplication equations?

I can represent the total number of items in equal groups by using repeated addition, unit form, or a multiplication equation. They can all be used to represent equal groups.

Exit Ticket 5 min

Provide up to 5 minutes for students to complete the Exit Ticket. It is possible to gather formative data even if some students do not complete every problem.

Sample Solutions

Expect to see varied solution paths. Accept accurate responses, reasonable explanations, and equivalent answers for all student work.

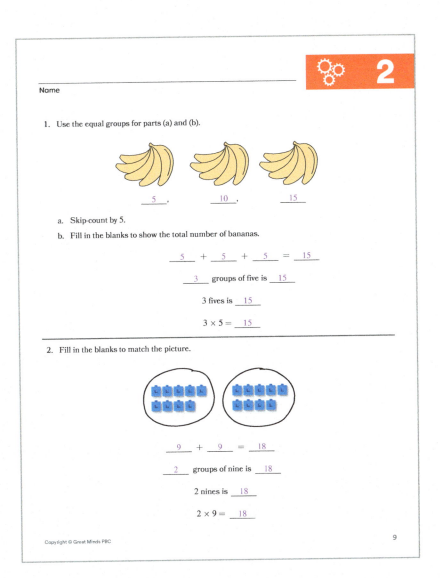

2

Name _____

1. Use the equal groups for parts (a) and (b).

_5_ , _10_ , _15_

a. Skip-count by 5.

b. Fill in the blanks to show the total number of bananas.

_5_ + _5_ + _5_ = _15_

_3_ groups of five is _15_

3 fives is _15_

$3 \times 5 =$ _15_

2. Fill in the blanks to match the picture.

_9_ + _9_ = _18_

_2_ groups of nine is _18_

2 nines is _18_

$2 \times 9 =$ _18_

9

3. The picture shows 2 groups of apples.

a. Does the picture show 2×3? Explain.

No. The expression 2×3 means 2 equal groups of 3. There are 2 groups, but they are not equal. The group on the right only has 2 apples.

b. Draw a picture to show $2 \times 3 = 6$.

Sample:

10 PROBLEM SET

4. Eva says, "I see six 3 times. We can multiply 3 × 6 to find the total number of eggs."

Do you agree with Eva? Explain.

I agree with Eva because there are 3 equal groups of 6 eggs. So, 3 × 6 can be used to find the total number of eggs.

3

Relate multiplication to the array model.

Name _____

✉ 3

Use the array for parts (a) and (b).

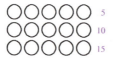

5
10
15

a. Skip-count by 5 down the side of the array.

b. Fill in the blanks to match the array.

___3___ groups of ___5___ is ___15___ .

___3___ fives is ___15___ .

___3___ × ___5___ = ___15___

Lesson at a Glance

Students build concrete arrays of fives and tens and relate them to equal groups, unit form, and multiplication to find the total. They relate the concrete array to a rekenrek and draw tape diagrams to represent equal groups. This lesson formalizes the term *product*.

Key Questions

- How does an array show equal groups?
- How does an array show each number in a multiplication equation?

Achievement Descriptor

3.Mod1.AD1 **Represent** a multiplication situation with a model and **convert** between several representations of multiplication. (3.OA.A.1)

Agenda

Fluency 10 min

Launch 5 min

Learn 35 min

- Groups of 5 in an Array
- Groups of 10 in an Array
- Arrays with a Rekenrek
- Tape Diagram
- Problem Set

Land 10 min

Materials

Teacher

- Interlocking cubes, 1 cm (20)
- 100-bead rekenrek

Students

- Interlocking cubes, 1 cm (20)
- Equal Groups (in the student book)

Lesson Preparation

- Prepare 2 sticks of 5 interlocking cubes in one color and 2 sticks of 5 interlocking cubes in a different color for each student and teacher.

- Print a copy of Equal Groups for use in the lesson.

- Tear out the Equal Groups page from the student books and place them inside personal whiteboards. Consider whether to prepare this material in advance or have students assemble it during the lesson.

- Review the Math Past resource to support delivery of Learn.

Fluency

Ready, Set, Add

Students find the total and say an addition equation to maintain fluency within 10 from grade 1.

Let's play Ready, Set, Add.

Have students form pairs and stand facing each other.

Model the action: Make a fist and shake it on each word as you say, "Ready, set, add." At "add," open your fist and hold up any number of fingers.

Tell students that they will make the same motion. At "add" they will show their partner any number of fingers. Consider doing a practice round with students.

Clarify the following directions:

- To show zero, show a closed fist at "add."
- Try to use different numbers each time to surprise your partner.

Each time partners show fingers, have them both say the total number of fingers. Then have each student say the addition equation, starting with the number of fingers on their own hand. See the sample dialogue under the photograph.

Circulate as students play the game to ensure that each student is trying a variety of numbers.

Partners A and B: "6"

Partner A: "$4 + 2 = 6$"

Partner B: "$2 + 4 = 6$"

Choral Response: Relating Multiplication Models

Students relate an equal groups picture with a unit of 5 or 10 to a repeated addition expression and unit form to develop an understanding of multiplication.

After asking each question, wait until most students raise their hands, and then signal for students to respond.

Raise your hand when you know the answer to each question. Wait for my signal to say the answer.

Show the picture of 2 groups of 5 apples.

What repeated addition expression represents this picture?

$5 + 5$

How do you represent the picture in unit form?

2 fives

Repeat the process with the following sequence:

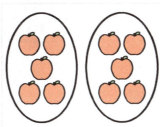

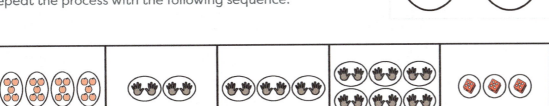

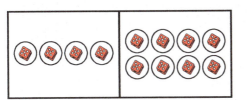

Teacher Note

Use hand signals to introduce a procedure for answering choral response questions. For example, cup your hand around your ear for *listen*, lift your finger to your temple for *think*, and raise your own hand to remind students to raise theirs.

Teach the procedure using general knowledge questions, such as:

- What grade are you in?

- What is the name of our school?

- What is your teacher's name?

Launch 5

Students consider ways to organize and count an unknown number of objects.

Show the picture of the bowl of cubes. Use the Math Chat routine to engage students in reasoning about efficient strategies they could use to organize and count to find the total number of cubes.

- Allow about 1 minute of silent time for students to think of ways to organize and count the cubes.

- Pair students to discuss their thinking. Circulate and listen as partners talk.

- Identify a few students to share their thinking. Purposely choose students with ideas that lead toward discussion of equal groups, efficiency, skip-counting by fives or tens, or that activate other knowledge from prior lessons.

- Ask the selected students to share their thinking with the class. Record their ideas and have them help facilitate discussion.

Transition to the next segment by framing the work.

> **Today, we will arrange equal groups in a different way and write equations to find the totals.**

Learn · 35

Groups of 5 in an Array

Materials—T/S: Cubes; S: Equal Groups

Students interpret an array and represent it as a multiplication equation.

Direct students to remove Equal Groups from their books and insert it into their whiteboards. Distribute 2 five-sticks of one color and 2 five-sticks of another color to each student. Display a set for the class to refer to. Make sure to leave space between the rows.

Arrange your sticks on your whiteboard like mine. Are your sticks organized into equal groups?

In grade 2, you arranged sticks into equal rows, like this with one under the other. When we use rows and columns to arrange equal groups, it is called an array.

Guide students to see the rows and columns in the array by tracing each row of 5 and column of 4. Refer to the sentence frames found on Equal Groups and guide students to complete the frames when prompted.

Look at your cubes. How many cubes are in each row?

How many times do you see a group of 5?

Let's skip-count by fives. Write the skip-count on your whiteboard down the side of the array.

5, 10, 15, 20

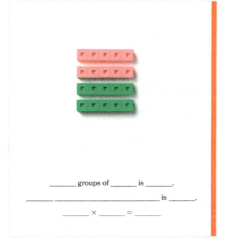

_____ groups of _____ is _____.
_____ _____ is _____.
_____ × _____ = _____

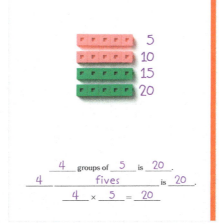

5
10
15
20

<u>4</u> groups of <u>5</u> is <u>20</u>.
<u>4</u> <u>fives</u> is <u>20</u>.
<u>4</u> × <u>5</u> = <u>20</u>

Record the skip-count down the side of the array.

Let's record: How many groups of 5 do we have?

What is the total of 4 groups of 5?

We can also describe the array in unit form. 4 fives is 20. Let's record: Write 4 fives is 20 on the second line.

Let's record: How can we write 4 fives is 20 as a multiplication equation?

$4 \times 5 = 20$

Ask students to relate each number in the multiplication equation back to the array. Ask them what each number in the multiplication equation represents in the array.

In multiplication, the answer, or total number, is called a product.

What is the product in the multiplication equation $4 \times 5 = 20$?

To pair the written term *product* with its meaning, use a different color marker to trace the 20 in the equation. Write *The product is* 20 below the equation.

Model putting together the 4 five-sticks to form a 4 by 5 array and ask students if it still shows 4 fives. Ask them what changed.

$4 \times 5 = 20$
The product is 20.

Does $4 \times 5 = 20$ still represent the array? Why or why not?

Yes, because the array did not change. We just put the five-sticks together to remove the spaces between the rows.

Invite students to turn and talk about how equal groups are represented in an array.

Promoting the Standards for Mathematical Practice

Students reason quantitatively and abstractly **(MP2)** as they discuss the array of cubes and corresponding multiplication, using two quantitative sentence frames (more concrete descriptions of the cubes) and two abstract sentence frames.

Ask the following questions to promote MP2:

- What does the product you found tell you about the cubes?

- What do the numbers in the first sentence frame tell you about the numbers in the last sentence frame?

Groups of 10 in an Array

Materials—T/S: Cubes

Students interpret an array and represent it as a multiplication equation.

Ask students to erase their whiteboards and combine their five-sticks to make ten-sticks, each with 5 cubes of one color and 5 cubes of another color.

Ask students what they notice about the relationship between fives and tens. Possible responses include:

I notice that 2 fives make 1 ten. $5 + 5 = 10$.

I notice that there are still 20 cubes. But now the cubes are in 2 rows of 10 instead of 4 rows of 5.

Repeat the sequence of instruction from the previous segment to complete the statements and equation.

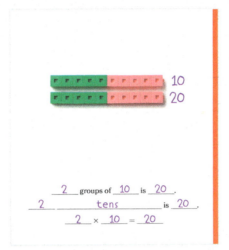

$\underline{2}$ groups of $\underline{10}$ is $\underline{20}$.

$\underline{2}$ $\underline{tens}$ is $\underline{20}$.

$\underline{2} \times \underline{10} = \underline{20}$

Teacher Note

Students progress through three stages of using arrays throughout module 1. First, they see familiar circle arrays from earlier grades, such as the rekenrek and 5-groups. Then students build arrays with square interlocking cubes and link them together to develop the big idea that all numbers can be units.

They initially build square arrays with gaps and then move to square arrays with no gaps. The square arrays without gaps support students in the transition to the area model later in module 4.

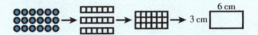

In module 1, when students are asked to draw arrays, prompt them to draw circles instead of squares. Students can draw circles more efficiently than they can squares, which require more precision.

Arrays with a Rekenrek

Materials—T: Rekenrek

Students interpret a pictorial array and represent it as multiplication.

Display a 100-bead rekenrek with only the top 3 rows showing. Direct students to the array of circles in their books.

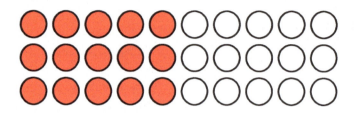

$$6 \times 5 = 30$$
$$3 \times 10 = 30$$

Math Past

Students might wonder why we use an × as a symbol for multiplication. Provide them with some fun facts about the history of the multiplication symbol from Olden ×.

Present some of the ways that × has been written in the past. Invite students to turn and talk to discuss each way and why they think the symbols were made. What similarities and differences do they notice among the symbols? Which symbol do they like best and why? Why is it that most people use the same symbol? Consider providing students with an opportunity to research other mathematical symbols and to share their findings.

Compare my rekenrek with the array shown in the book. What is different about them?

Yours has red beads and white beads, but the drawing has only white beads.

Fives are easier to see on the rekenrek than in the array.

Ask students how many red beads are in each row and how many white beads are in each row. Have students shade the first 5 circles in each row of their books to match the rekenrek.

Look at your array. How many times do you see groups of 5?

I see five 6 times.

Where do you see the 6 fives?

Ask students to write a multiplication equation in their books to describe the fives in the array.

What multiplication equation describes the 6 fives?

$6 \times 5 = 30$

Let's count by fives the math way to check our work. Each finger represents 5. Count with me.

5, 10, 15, 20, 25, 30

Ask students to look at their array again and describe how many times they see groups of 10. Guide students to explain where they see 3 tens in the array and have them write a related multiplication equation in their books.

What multiplication equation did you write?

$3 \times 10 = 30$

Let's count by tens the math way to check our work. This time when we count, each finger represents a unit of ten. Count with me.

10, 20, 30

What do you notice about 3×10 and 6×5?

The product of each is 30, so they are equal.

It's the same array. We saw groups of 5 inside the groups of 10.

Invite students to turn and talk about how the same array can represent groups of 5 and groups of 10.

Tape Diagram

Students relate tape diagrams to equal groups.

Use the following sequence to interactively model drawing a tape diagram.

Watch as I draw 3 tens using a tape diagram.

Model drawing the outline of the tape diagram.

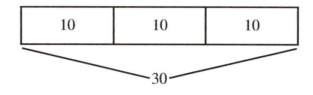

How many equal groups are in 3 tens?

Model partitioning the tape diagram into three equal groups.

How many are in each group?

Label each part of the tape diagram as 10.

What is the total, or product, of 3 tens?

Label the product on the tape diagram. Invite students to turn and talk about where they see the multiplication equation $3 \times 10 = 30$ in the tape diagram.

Problem Set

Differentiate the set by selecting problems for students to finish independently within the timeframe. Problems are organized from simple to complex.

Debrief 5 min

Objective: Relate multiplication to the array model.

Engage the class in discussion about the relationship between multiplication and the array model.

How does an array show equal groups?

The equal groups are the rows in the array.

How does an array show each number in a multiplication equation?

An array has equal groups and you can see the total.

The array shows the equal groups and you can see how many groups there are.

What is the difference between an equal-groups picture and an array?

They both show equal groups and the total, but they are organized differently.

Exit Ticket 5 min

Provide up to 5 minutes for students to complete the Exit Ticket. It is possible to gather formative data even if some students do not complete every problem.

Sample Solutions

Expect to see varied solution paths. Accept accurate responses, reasonable explanations, and equivalent answers for all student work.

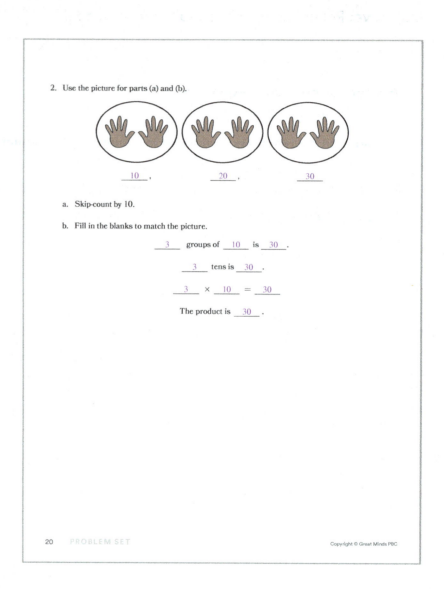

Name

1. Use the picture for parts (a) and (b).

    __5__ , __10__ , __15__ , __20__ , __25__ , __30__

    a. Skip-count by 5.

    b. Fill in the blanks to match the picture.

    __6__ groups of 5 is __30__ .

    __6__ fives is __30__ .

    __6__ × __5__ = __30__

2. Use the picture for parts (a) and (b).

    __10__ , __20__ , __30__

    a. Skip-count by 10.

    b. Fill in the blanks to match the picture.

    __3__ groups of __10__ is __30__ .

    __3__ tens is __30__ .

    __3__ × __10__ = __30__

    The product is __30__ .

19

20   PROBLEM SET

3. Use the array for parts (a)–(c).

⬤⬤⬤⬤⬤ 5
⬤⬤⬤⬤⬤ 10
⬤⬤⬤⬤⬤ 15
⬤⬤⬤⬤⬤ 20
⬤⬤⬤⬤⬤ 25
◯◯◯◯◯ 30
◯◯◯◯◯ 35
◯◯◯◯◯ 40

a. Skip-count by 5 down the side of the array.

b. Fill in the blanks to match the array.

  __8__ fives is __40__ .

  __8__ × __5__ = ⟨40⟩

c. Circle the product in the equation.

4. Use the array for parts (a)–(c).

a. Skip-count by 10 down the side of the array.

b. Fill in the blanks to match the array.

  __4__ tens is __40__ .

  __4__ × __10__ = ⟨40⟩

c. Circle the product in the equation.

5. What do you notice about 8 × 5 and 4 × 10?

Each expression is equal to 40, so they have the same value.

8 × 5 = 40 and 4 × 10 = 40

4

Interpret the meaning of factors as number of groups or number in each group.

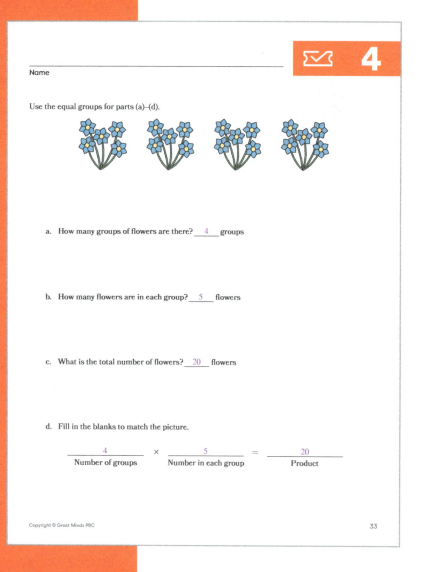

Lesson at a Glance

Students relate equal groups, arrays, and tape diagrams to multiplication equations. They interpret the meaning of factors in equations as either the number of groups or the number in each group. This lesson introduces the term *factor*.

Key Questions

- What do the factors represent in multiplication equations?
- How can the same multiplication equation represent an equal group picture and an array?

Achievement Descriptor

3.Mod1.AD1 **Represent** a multiplication situation with a model and **convert** between several representations of multiplication. (3.OA.A.1)

Agenda

Fluency 10 min

Launch 5 min

Learn 35 min

- Relate Equal Groups to Arrays
- Equal Groups, Arrays, and Equations
- Arrays, Tape Diagrams, and Equations
- Problem Set

Land 10 min

Materials

Teacher

- 100-bead rekenrek
- Interlocking cubes, 1 cm (15)

Students

- None

Lesson Preparation

Gather 15 interlocking cubes in one color.

Fluency

Counting on the Rekenrek by Fives

Materials—T: Rekenrek

Students count by fives in unit and standard form to develop an understanding of multiplication.

Show students the rekenrek. Start with all the beads to the right side.

> **Say how many beads there are as I slide them over.**

Slide 5 red beads in the top row all at once to the left side.

> 5

> **The unit is 5. In unit form, we say 1 five. Say 5 in unit form.**

> 1 five

Slide the 5 red beads in the next row all at once to the left side.

> **How many beads are there now? Say it in unit form.**

> 2 fives

Continue sliding over 5 red beads in each row all at once as students count.

> 3 fives, 4 fives, 5 fives, 6 fives, 7 fives, 8 fives, 9 fives, 10 fives

Slide all the beads back to the right side.

> **Now let's practice counting in standard form. Say how many beads there are as I slide them over.**

> **Let's start at 0. Ready?**

Slide over 5 beads in each row all at once as students count.

> 0, 5, 10, 15, 20, 25, 30, 35, 40, 45, 50

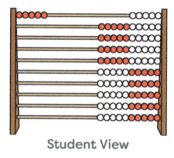

Student View

Teacher Note

Take care not to count or mouth the words along with students. Students may learn to mimic the teacher rather than focus on number order.

Counting the Math Way by Fives

Students construct a number line with their hands while counting aloud by fives to develop a strategy for multiplying.

Let's count the math way. Each finger represents 5.

Face the students and instruct them to mirror you. Show a fist with your right hand, palm facing out.

Show me your left hand. Make a fist like me. That's 0.

Now, raise your right pinkie.

Now, show me your left pinkie. That's 5.

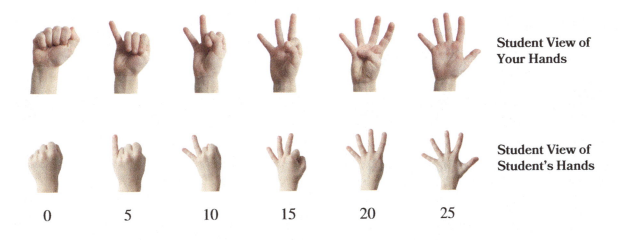

Student View of Your Hands

Student View of Student's Hands

0   5   10   15   20   25

Let's put up the very next finger.

Raise your right ring finger. Students raise their left ring finger.

That's 10.

Put up the next finger. 15.

Now that students understand the routine, switch to having them say the count as they show fingers. Guide students to continue counting the math way by fives to 50, then back down to 0.

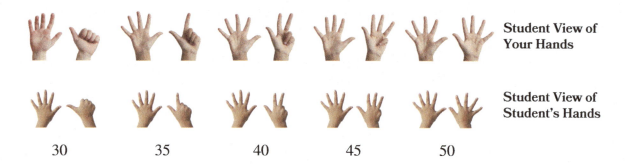

Student View of
Your Hands

Student View of
Student's Hands

30   35   40   45   50

Choral Response: Relating Multiplication Models

Students relate an equal groups picture, array, or tape diagram with a unit of 5 or 10 to a repeated addition expression, unit form, and multiplication equation to develop an understanding of multiplication.

After asking each question, wait until most students raise their hands, and then signal for students to respond.

Raise your hand when you know the answer to each question. Wait for my signal to say the answer.

Show the picture of 2 groups of 5 apples.

What repeated addition expression represents this picture?

$5 + 5$

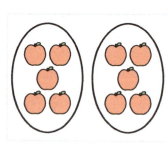

How do you represent the picture in unit form?

2 fives

What multiplication equation represents this picture?

$2 \times 5 = 10$

Repeat the process with the following sequence:

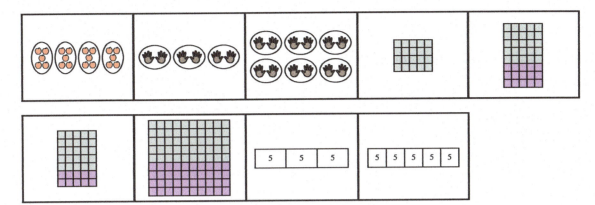

Teacher Note

Arrays in the sequence of pictures with more than 5 rows are shaded to support students in quickly determining the number of rows in each array without having to count each one.

Launch 5

Students learn the term *factor* and relate it to a multiplication equation.

Ask students to stand and, without talking, organize themselves into groups of 4.

When they have finished, encourage students to verify that the groups are equal, and ask the class the following questions.

How many equal groups did we make?

How many students are in each group?

How many total students are there?

Write the sentence frame and equation frame as shown.

_____ groups of 4 is _____.

_____ × 4 = _____

Teacher Note

If you can't arrange your class into equal groups of 4, adjust the directions for the beginning of Launch to make equal groups of a different size. If it's not possible to make equal groups, consider having an adult, a teddy bear, or a class pet stand in to adjust the size of the groups. Avoid having students make 4 groups of 4.

Direct students to complete the sentence frame and equation orally with their group. Then restate the completed sentence frame and equation for the class.

Which number in our equation is the product?

In a multiplication equation, the two numbers we multiply together are called factors.

Circle the factors in the equation and refer to them as you discuss the term *factor*.

In this example, _____ is the first factor and 4 is the second factor.

Consider creating an anchor chart for the terms *product* and *factor* along with a pictorial example of each to support students in their precision using each term.

$$\underset{\text{factor}}{_____} \times \underset{\text{factor}}{_____} = \underset{\text{product}}{_____}$$

Transition to the next segment by framing the work.

Today, we will learn what the factors in a multiplication expression represent.

Learn ⏱35

Relate Equal Groups to Arrays

Materials—T/S: Cubes

Students draw an array to represent groups of concrete objects and interpret the meaning of factors.

Display 3 groups of 5 cubes.

Ask students how many groups there are and how many cubes are in each group. Have the class skip-count by fives to find the total.

Rearrange the groups to make an array with 3 rows of 5. Move one group of cubes at a time.

How did I organize the equal groups?

You put the equal groups into rows.

You made an array.

The equal groups are the rows in the array. Do I still have 3 equal groups?

Do I still have 5 cubes in each group?

Do I still have 15 as the total?

Direct students to problem 1 in their books. Ask students to draw an array to show 3 rows of 5, using circles to represent the cubes.

1. Draw an array to show 3 rows of 5.

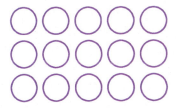

$3 \times 5 = 15$

When most students have finished, write $3 \times 5 = 15$.

15 is the product. What do we call 3 and 5?

Factors

What does the 3 represent in the array?

3 rows

What does the 5 represent in the array?

The 5 circles in each row

Teacher Note

The digital interactive Groups and Arrays helps students visualize the connection between the properties of equal groups and arrays.

Consider allowing students to experiment with the tool individually or demonstrating the activity for the whole class.

UDL: Action & Expression

Consider providing resources to support organization and alternatives that minimize fine-motor demands when students are required to draw arrays. Examples include the following:

- Provide lined paper to help draw circles in even rows.

- Provide stamp markers to create circles of the same size.

- Provide physical manipulatives, such as circle-shaped counters as an alternative to drawing.

Tell students that one way we use factors to write a multiplication equation is to have the first factor represent the number of groups. The second factor represents the number in each group.

Equal Groups, Arrays, and Equations

Students draw arrays to represent pictures of equal groups and interpret the meaning of factors.

Pair students and direct them to complete problem 2.

2. Use the equal groups for parts (a)–(d).

a. There are ____6____ groups of balloons.

There are ____5____ balloons in each group.

There is a total of ____30____ balloons.

b. Complete the equation to describe the equal groups.

$$\underset{\text{Number of groups}}{\underline{\qquad 6 \qquad}} \times \underset{\text{Number in each group}}{\underline{\qquad 5 \qquad}} = \underset{\text{Product}}{\underline{\qquad 30 \qquad}}$$

c. Draw an array to show the total number of balloons. Make each equal group a row.

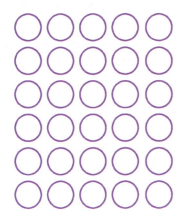

d. Complete the equation to describe the array.

$$\underline{\qquad 6 \qquad} \times \underline{\qquad 5 \qquad} = \underline{\qquad 30 \qquad}$$

Number of rows   Number in each row   Product

As students work, consider asking partners the following questions:

- Where are the 6 groups of balloons in the array?
- Where are the 5 balloons in each group in the array?
- What part of the equation represents the number of rows in the array?
- What part of the equation represents the number in each row in the array?
- What part of the equation represents the total, or product?

After students finish problem 2, show the picture of 3 plates of crackers.

Ask the class the following questions:

How many groups of crackers are there?

How many crackers are in each group?

What is the total number of crackers?

Ask students to take out their whiteboards and draw the equal groups of crackers as an array. Then have them write a matching multiplication equation.

Arrays, Tape Diagrams, and Equations

Students relate an array and a tape diagram to the factors in a multiplication equation.

How would we draw an array model to show 7 groups of 5?

We would draw 7 rows of 5 circles.

Direct students to problem 3 and ask them to draw the array and answer the questions.

3. Draw an array to show 7 groups of 5. Make each group a row.

○ ○ ○ ○ ○
○ ○ ○ ○ ○
○ ○ ○ ○ ○
○ ○ ○ ○ ○
○ ○ ○ ○ ○
○ ○ ○ ○ ○
○ ○ ○ ○ ○

a. What factor shows the number of groups? ___7___

b. What factor shows the number in each group? ___35___

c. What is the product? ___5___

d. Complete the equation to describe the array.

$$\underset{\text{Number of rows}}{\underline{\hspace{1cm}7\hspace{1cm}}} \times \underset{\text{Number in each row}}{\underline{\hspace{1cm}5\hspace{1cm}}} = \underset{\text{Product}}{\underline{\hspace{1cm}35\hspace{1cm}}}$$

Select a student work sample to display and use it to lead a discussion about the meaning of factors. Consider asking the following questions:

- How does the array show the number of groups?
- How does the array show the number in each group?
- How could we use a tape diagram to show 7 groups of 5?

Model drawing the tape diagram.

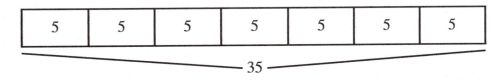

What multiplication equation represents the tape diagram? The first factor should be 7.

Write $7 \times 5 = 35$.

Invite students to turn and talk about how the factors and product relate to the tape diagram.

How are the array and the tape diagram similar?

They both have a total of 35.

The array has 7 rows, and the tape diagram has 7 equal groups.

There are 5 circles in each row of the array, and 5 is in each group of the tape diagram.

Invite students to turn and talk about how arrays and tape diagrams can represent equal groups.

Problem Set

Differentiate the set by selecting problems for students to finish independently within the timeframe. Problems are organized from simple to complex.

Teacher Note

Help students recognize the words *complete* and *equation* in the Problem Set. Consider providing extra support, such as reading the problems aloud and underlining the terms.

Land

Debrief 5 min

Objective: Interpret the meaning of factors as number of groups or number in each group.

Use the following prompts to guide a discussion about the meaning of each factor in an equation and how equations relate to pictorial models such as equal groups, arrays, and tape diagrams.

What do the factors represent in multiplication equations?

The factors represent the number of groups and the number in each group.

They represent the number of circles in each row.

How can the same multiplication equation represent an equal-group picture and an array?

Both the equal groups and the array represent 4 fives and have a total of 20.

The number of rows and groups are the same. So are the number in each row and the number in each group.

Exit Ticket 5 min

Provide up to 5 minutes for students to complete the Exit Ticket. It is possible to gather formative data even if some students do not complete every problem.

Sample Solutions

Expect to see varied solution paths. Accept accurate responses, reasonable explanations, and equivalent answers for all student work.

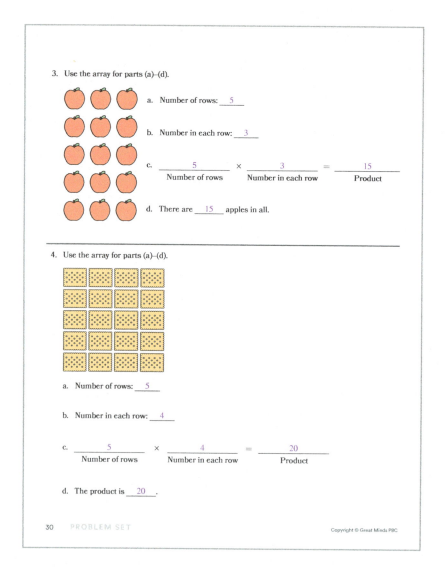

Name _____ 🛠 **4**

1. Use the equal groups for parts (a)–(c).

   a. How many groups of cars are there? __4__ groups

   b. How many cars are in each group? __5__ cars are in each group

   c. Complete the equation.

   __4__ × __5__ = __20__
   Number of groups   Number in each group   Product

2. Use the array for parts (a)–(c).

   a. How many rows of cars are there? __4__ rows of cars

   b. How many cars are in each row? __5__ cars are in each row

   c. Complete the equation.

   __4__ × __5__ = __20__
   Number of rows   Number in each row   Product

3. Use the array for parts (a)–(d).

   a. Number of rows: __5__

   b. Number in each row: __3__

   c. __5__ × __3__ = __15__
   Number of rows   Number in each row   Product

   d. There are __15__ apples in all.

4. Use the array for parts (a)–(d).

   a. Number of rows: __5__

   b. Number in each row: __4__

   c. __5__ × __4__ = __20__
   Number of rows   Number in each row   Product

   d. The product is __20__.

5. David drew an array and wrote an equation.

$$6 \times 5 = 30$$

a. Which factor tells the number of rows? __6__

b. Which factor tells the number in each row? __5__

c. What is the product? __30__

5

Represent and solve multiplication word problems by using drawings and equations.

Name

Use the Read–Draw–Write process to solve the problem.

Mr. Endo has 5 boxes of crayons.

There are 8 crayons in each box.

How many crayons does Mr. Endo have?

Sample:

$$8 + 8 + 8 + 8 + 8 = 40$$
$$5 \times 8 = 40$$

Mr. Endo has 40 crayons.

Lesson at a Glance

Students select strategies to represent and solve multiplication word problems by using drawings and equations. A video provides context for the word problems. After working independently to solve the problems, students share their work to compare and connect various representations and strategies.

Key Questions

- How is a tape diagram a useful model to use when solving multiplication word problems?
- How do you decide which model to use when solving a multiplication word problem?

Achievement Descriptor

3.Mod1.AD3 **Solve** one-step word problems by using multiplication and division within 100, involving factors and divisors 2–5 and 10. (3.OA.A.3)

Agenda

Fluency 5 min

Launch 5 min

Learn 40 min

- Equal Groups Word Problem
- Equal Groups: Share, Compare, and Connect
- Array Word Problem
- Array: Share, Compare, and Connect
- Problem Set

Land 10 min

Materials

Teacher

- None

Students

- Interlocking cubes, 1 cm (45)

Lesson Preparation

None

Fluency 5

Choral Response: Relating Multiplication Models

Students relate an equal groups picture, an array, or a tape diagram with a unit of **5** or **10** to a repeated addition expression, a unit form, and a multiplication equation to develop an understanding of multiplication.

After asking each question, wait until most students raise their hands, and then signal for students to respond.

Raise your hand when you know the answer to each question. Wait for my signal to say the answer.

Display the picture of 3 dice.

What repeated addition expression represents this picture?

$5 + 5 + 5$

How do you represent the picture in unit form?

3 fives

What multiplication equation represents this picture?

$3 \times 5 = 15$

Repeat the process with the following sequence:

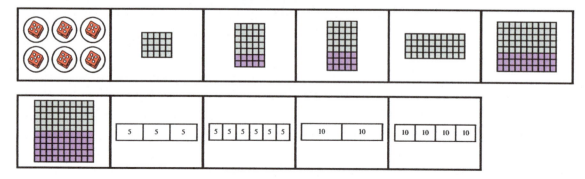

Teacher Note

Arrays in the sequence of pictures with more than 5 rows are shaded to support students in quickly determining the number of rows in each array without having to count each one.

Students work together to determine which picture accurately represents a given multiplication scenario.

Display the picture of 3 card arrangements and read the following problem aloud to the class.

> Casey plays a card matching game. To start the game, Casey puts her cards in 5 rows with 4 cards in each row. How many total cards does Casey use to play the matching game?

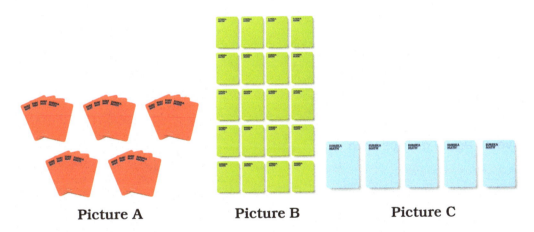

Picture A   Picture B   Picture C

Invite students to turn and talk about whether picture A, picture B, or picture C accurately represents the problem about Casey's cards. Have students turn and talk to respond to the following questions:

- How are the cards represented in picture A? Picture B? Picture C?

- Where do you see 5 in picture A? Picture B? Picture C?

- Where do you see 4 in picture A? Picture B? Picture C?

- How are the pictures similar? How are they different?

Ask which picture represents the problem about Casey's cards.

How many total cards does Casey use to play the matching game?

Which picture shows a total of 20 cards?

Since all three pictures show a total of 20 cards, any picture can be used to find the total.

If all three pictures show the same total, why did you choose picture B?

All three pictures show 5 fours, just in different ways. Since the problem said Casey makes 5 rows of 4, we chose picture B because it is an array with 5 rows of 4.

Transition to the next segment by framing the work.

Today, we will solve word problems and share and compare our strategies.

Learn

Equal Groups Word Problem

Students collect information from a video and solve an *equal groups with unknown product* word problem.

Play part 1 of the Amusement Park video. If necessary, replay the video and ask students to note any details.

Give students 1 minute to turn and talk about what they noticed.

Engage students in a brief conversation about the video. Discuss student observations and any relevant questions they have. Guide the conversation to problem 1. Consider the following possible sequence of questions to ask students:

What do you notice?

I saw people getting on a roller coaster.

The roller coaster had 10 cars.

There were 3 people in each car.

Teacher Note

This is the first use of a context video. It is shown before a related word problem to build familiarity and engagement with the context. It also allows students to visualize and discuss the situation before being asked to interpret it mathematically.

What do you wonder?

How many people fit on the roller coaster?

There are many mathematical questions we could ask. Let's use what we saw in the video to help us understand and solve a word problem.

Direct students to problem 1, and chorally read the problem with the class. Have students work independently to use the Read–Draw–Write process to solve the problem. Provide materials such as interlocking cubes for student use. Encourage students to self-select their tools and strategies.

Use the Read–Draw–Write process to solve the problem.

1. A roller coaster has 10 cars.

   There are 3 people in each car.

   How many people are on the roller coaster?

   $10 \times 3 = 30$

   There are 30 people on the roller coaster.

Circulate and observe student strategies. Select two or three students to share in the next segment. Look for work samples that help advance the lesson's objective of using different multiplication models, such as equal groups, arrays, and tape diagrams.

The following student work samples demonstrate using equal groups and a tape diagram to represent multiplication.

Copyright © Great Minds PBC

UDL: Action & Expression

Some students may benefit from using cubes to model the problem.

Prompt students who use repeated addition to transition to multiplication by using unit form with the sentence frame:

I see _____ threes.

Then follow up by asking: How many times do you see 3?

Use the equation frame:
_____ × _____ = _____.

Promoting the Standards for Mathematical Practice

Students model with mathematics (MP4) as they iteratively create a drawing and equation to represent and solve a word problem (i.e., the Read–Draw–Write process).

Ask the following questions to promote MP4:

- What can you draw to help you understand the roller coaster problem?

- What kind of math could you use to represent your model?

- What key pieces of information from the roller coaster problem should be in your model and your equation?

Equal Groups

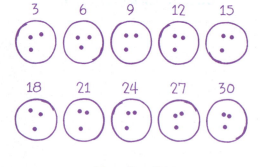

10 x 3 = 30

There are 30 people
on the roller coaster.

Tape Diagram

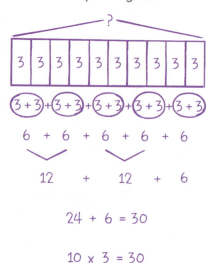

24 + 6 = 30

10 x 3 = 30

There are 30 people
on the roller coaster.

Teacher Note

The sample student work shows common responses. Look for similar work from your students, and encourage authentic classroom conversations about the key concepts.

If your students do not produce similar work, choose one or two pieces of their work to share, and highlight how it shows movement toward the goal of this lesson. Then select one work sample from the lesson that works best to advance student thinking. Consider presenting the work by saying, "This is how another student solved the problem. What do you think this student did?"

Equal Groups Word Problem: Share, Compare, and Connect

Students share solutions for problem 1 and reason about their connections.

Gather the class, and invite the students you identified in the previous segment to share their solutions one at a time. Consider intentionally ordering shared student work from a representational model, such as an equal groups drawing, to a more abstract model, such as a tape diagram.

As each student shares, ask questions to elicit their thinking and clarify the model used to represent the problem. Ask the class questions to make connections between the different solutions and their own work. Encourage students to ask questions of their own.

Teacher Note

Throughout module 1, students are encouraged to complete tape diagrams by labeling all components; the number of groups, the number in each group, and the product. This supports their understanding of the problems and validates their solutions.

In module 3, as the size of factors increases and students gain fluency with their multiplication facts, they transition to using a symbol to identify the unknown.

Equal Groups (Luke's Way)

What did Luke do in his drawing?

He drew 10 equal groups to represent the 10 roller coaster cars. Then he drew 3 dots in each group to represent the 3 people in each car.

What strategy did Luke use?

He skip-counted by threes like this: 3, 6, 9, 12, …, 30.

Why did you decide to represent the problem with equal groups?

When I read the problem, I thought about equal groups because I pictured 10 cars with 3 people in each group. That's the same as 10 groups with 3 in each group.

What multiplication equation represents the problem? Why?

$10 \times 3 = 30$ because the number of groups is 10, the number in each group is 3, and the total number of people is 30.

Invite students to turn and talk about the similarities and differences between Luke's work and their work.

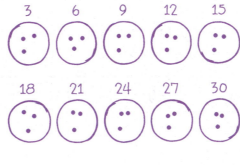

$10 \times 3 = 30$

There are 30 people on the roller coaster.

Language Support

Students may need support in sharing their thinking using detail and academic language. Consider modeling a think aloud for a solution strategy.

Also consider posting sentence frames for students to reference until they are more comfortable sharing their thinking in a way that allows other students to follow their solution sequence. Sentence frames could include:

- First, I drew a _____ to represent the problem.

- I chose to draw a _____ because _____.

- I wrote the equation _____ because _____.

- My strategy to find the total was to _____.

- My strategy is similar to/different from _____ because _____.

Tape Diagram (Eva's Way)

What did Eva do in her drawing?

She drew a tape diagram with 10 parts to represent the 10 roller coaster cars. She wrote the number 3 in each part to represent the 3 people in each car. She wrote a question mark to show that she needed to find the total number of people.

What strategy did Eva use?

She used doubles facts and added one extra 6 at the end to get 30.

Why did you decide to represent the problem with a tape diagram?

I used a tape diagram because I thought about equal groups. The tape diagram shows the equal groups.

What multiplication equation represents the problem? Why?

$10 \times 3 = 30$ because 10 represents the 10 roller coaster cars, 3 represents the 3 people in each car, and 30 represents the total number of people.

How is Luke and Eva's thinking similar?

They both thought about equal groups.

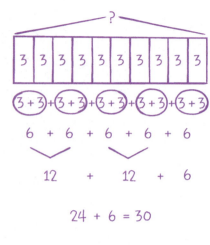

$$24 + 6 = 30$$

$$10 \times 3 = 30$$

There are 30 people on the roller coaster.

Invite students to turn and talk about the similarities and differences between Eva's work and their work.

Array Word Problem

Students collect information from a video and solve an *array with unknown product* word problem.

Play part 2 of the Amusement Park video. If necessary, replay the video, and ask students to note any details.

Give students 1 minute to turn and talk about what they noticed.

Engage students in a brief conversation about the video. Discuss student observations and any relevant questions they have. Guide the conversation to problem 2. Consider the sequence of questions used in part 1.

Direct students to problem 2. Chorally read the problem with the class. Prompt students to use the Read–Draw–Write process to solve the problem. Provide materials such as interlocking cubes for student use. Encourage students to self-select their tools and strategies.

> Use the Read–Draw–Write process to solve the problem.
>
> 2. The swinging ship ride has 9 rows of seats.
>
>   Each row has 5 seats.
>
>   How many seats are on the swinging ship ride?
>
> $9 \times 5 = 45$
>
>   There are 45 seats on the swinging ship ride.

Circulate and observe student strategies. Select two or three students to share in the next segment. Look for work samples that help advance the lesson's objective of using different multiplication models, such as equal groups, arrays, and tape diagrams.

The student work samples demonstrate using an array and a tape diagram to represent multiplication.

Array

○○○○○ 5
○○○○○ 10
○○○○○ 15
○○○○○ 20
○○○○○ 25
○○○○○ 30
○○○○○ 35
○○○○○ 40
○○○○○ 45

9 x 5 = 45

There are 45 seats on the swinging ship ride.

Tape Diagram

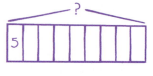

9 x 5 = 45

There are 45 seats on the swinging ship ride.

Array Word Problem: Share, Compare, and Connect

Students share solutions for problem 2 and reason about their connections.

Gather the class, and invite the students you identified in the previous segment to share their solutions one at a time. Consider intentionally ordering shared student work from a representational model (array) to a more abstract model (tape diagram).

As each student shares, ask questions to elicit their thinking and clarify the model used to represent the problem. Ask the class questions to make connections between the different solutions and their own work. Encourage students to ask questions of their own.

Array (Mia's Way)

What did Mia do in her drawing?

She drew 5 circles in each row to represent the 5 seats in each row. She continued to draw rows of 5 until she had 9 rows because there are 9 rows of seats on the ride.

What strategy did Mia use to solve the problem?

She skip-counted by fives. I can see it at the end of each row.

What is useful about representing the problem with an array?

It helps me imagine all the parts of the problem.

What equation represents the problem? Why?

$9 \times 5 = 45$ because the number of rows is 9, the number in each row is 5, and the total is 45.

○○○○○ 5
○○○○○ 10
○○○○○ 15
○○○○○ 20
○○○○○ 25
○○○○○ 30
○○○○○ 35
○○○○○ 40
○○○○○ 45

$9 \times 5 = 45$

There are 45 seats on the swinging ship ride.

Teacher Note
Students may represent their work differently than the sample, such as by drawing an array with 5 rows of 9 circles and writing the equation $5 \times 9 = 45$. Accept other answers as long as they accurately represent the problem.

Invite students to turn and talk about the similarities and differences between Mia's work and their work.

Tape Diagram (Pablo's Way)

What did Pablo do in his drawing?

He drew a tape diagram with 9 parts to represent the 9 rows of seats. He wrote 5 in one of the parts because he knew each row has the same number of seats, so he didn't need to keep writing 5 in every part. He drew a question mark to show that he needed to find the total number of seats.

How is the total number of seats represented in the tape diagram?

The total number of seats is represented by the 9 parts in the tape diagram.

What strategy did Pablo use to solve the problem?

I think he skip-counted by fives 9 times.

What is useful about representing the problem with a tape diagram?

It is quicker to draw a tape diagram than an array.

What equation represents the problem? Why?

$9 \times 5 = 45$ because there are 9 rows with 5 in each row, which makes a total of 45.

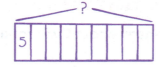

$9 \times 5 = 45$

There are 45 seats on the swinging ship ride.

Invite students to turn and talk about the similarities and differences between

• Mia's work and Pablo's work, and

• Pablo's work and their work.

To support the discussion in Land, save two or three student solutions that use tape diagrams.

Problem Set

Differentiate the set by selecting problems for students to finish independently within the timeframe. Problems are organized from simple to complex.

Debrief 5 min

Objective: Represent and solve multiplication word problems by using drawings and equations.

Use the following prompts to guide a discussion about the usefulness of equal groups, arrays, and tape diagrams to represent different multiplication scenarios. To support this discussion, display selected student work that uses tape diagrams.

How is the tape diagram a useful model to use when solving multiplication word problems?

Tape diagrams help me make sense of the problem.

I can see the equal groups and the number in each group with a tape diagram.

How can you use a tape diagram to find the total in a multiplication problem?

You can skip-count.

You can add all the parts together.

You can use repeated addition or doubles to add all the parts together.

How do you decide which model to use when solving a multiplication word problem?

I draw an equal groups picture or an array so I can count to find the total.

I read the problem and picture what it is about. Like in problem 2, it said 9 rows with 5 seats in each row. The word *row* helped me picture an array.

I like to draw tape diagrams because they can show equal groups or equal rows.

Exit Ticket 5 min

Provide up to 5 minutes for students to complete the Exit Ticket. It is possible to gather formative data even if some students do not complete every problem.

Sample Solutions

Expect to see varied solution paths. Accept accurate responses, reasonable explanations, and equivalent answers for all student work.

5

Name _____

Use the Read–Draw–Write process to solve each problem.

1. Oka has 4 baskets.
   Each basket has 5 bagels.
   How many bagels does Oka have?

   $4 \times 5 = 20$

   Oka has ___20___ bagels.

2. A muffin pan has 5 rows.
   Each row has 3 muffins.
   How many muffins are in the pan?

   $5 \times 3 = 15$

   There are ___15___ muffins in the pan.

3. Mr. Lopez has 10 packs of cups.
   Each pack has 6 cups.
   How many cups does Mr. Lopez have?

   $10 \times 6 = 60$

   Mr. Lopez has 60 cups.

4. Ivan has 6 packs of tomatoes.
   Each pack has 5 tomatoes.
   How many tomatoes does Ivan have?

   $6 \times 5 = 30$

   Ivan has 30 tomatoes.

5. A garden has 4 rows of beets.
   Each row has 10 beets.
   How many beets are in the garden?

   $4 \times 10 = 40$

   There are 40 beets in the garden.

6. Mrs. Smith has 10 boxes of books.
   Each box has 10 books.
   How many books does Mrs. Smith have?

   $10 \times 10 = 100$

   Mrs. Smith has 100 books.

37

38 PROBLEM SET

Topic B
Conceptual Understanding of Division

In topic B, students develop a conceptual understanding of division, following a sequence similar to topic A's development of multiplication. Throughout the topic, two interpretations of division are presented in tandem. Sometimes the number of groups is known (partitive division), and sometimes the number in each group is known (measurement division). Students do not use the terminology of partitive and measurement division, but they do identify what is known and unknown in each situation. They develop an understanding that the unknown affects the choice of modeling, the solution path, and the meaning of the solution. Problems are contextualized so students gain experience identifying whether a number represents the number in each group, the number of groups, or the total.

Students transition from representing division situations with equal groups models to representing them with arrays and equations by using the ÷ symbol. They make the connection between the rows and columns of an array and the number of groups and the number in each group in the problem situation. They relate these representations to the quantities in the division equation.

When solving real-world problems, students determine whether an equal groups model or an array model best represents the situation. They create drawings to represent the situations and then use their drawings to write equations and determine an appropriate solution path before solving the problem.

In topic C, students extend their work from topic A, applying their understanding of arrays as equal groups to the commutative property of multiplication and to the distributive property. Topic C focuses on the units of 2, 3, 4, 5, and 10.

Progression of Lessons

Lesson 6

Explore measurement and partitive division by modeling concretely and drawing.

My strategy for equal sharing when I know the total and the number in each group is different from my strategy when I know the total and the number of groups.

Lesson 7

Model measurement and partitive division by drawing equal groups.

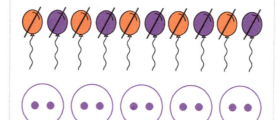

There are 5 equal groups.

There are 4 pieces of candy in each group.

I draw equal groups pictures to represent division situations. Sometimes I know the total and the number in each group, and sometimes I know the total and the number of groups.

Lesson 8

Model measurement and partitive division by drawing arrays.

Gabe's Seashells   Gabe's Rocks

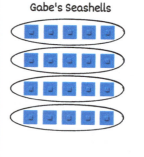

20	÷	5	=	4
total		number in each row		number of rows

15	÷	5	=	3
total		number of rows		number in each row

I use arrays and equations to represent division situations. The ÷ symbol means 'divided by' and I use the symbol when I write division expressions and equations.

Lesson 9

Represent and solve division word problems using drawings and equations.

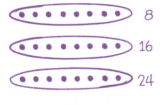

8

16

24

$24 ÷ 8 = 3$

Miss Wong makes
3 rows of pictures.

There are many ways to represent and solve division word problems using drawings and equations. I select strategies that I've learned that make sense to me.

6

Explore measurement and partitive division by modeling concretely and drawing.

Name _____

✉ 6

There are 12 balloons.

a. Circle groups of 4 balloons.

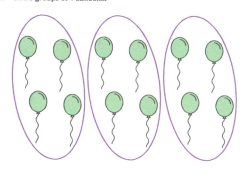

b. Fill in the blanks to match the picture.

The number in each group is ___4___.

The number of equal groups is ___3___.

Lesson at a Glance

Students use concrete models in equal-sharing division situations where the total and either the number of groups or the number in each group are known. Students describe how their modeling changes based on what is known.

Key Questions

- How can equal sharing be done in different ways?
- Why is it helpful to think about what the numbers in a division problem represent?

Achievement Descriptors

3.Mod1.AD2 Represent a division situation with a model and **convert** between several representations of division. (3.OA.A.2)

Agenda

Fluency 10 min

Launch 5 min

Learn 35 min

- Equally Share Ten Crackers
- Equally Share Twenty Crackers
- Problem Set

Land 10 min

Materials

Teacher

- None

Students

- Paper plates (5 per student pair)
- Crackers (20 per student pair)

Lesson Preparation

None

Fluency 🔟

Choral Response: Equal Parts

Students identify and describe equal parts of a shape to maintain geometry concepts from grade 2.

After asking each question, wait until most students raise their hands, and then signal for students to respond.

Raise your hand when you know the answer to each question. Wait for my signal to say the answer.

Display the circle.

What is the name of this shape?

Circle

This is 1 whole circle.

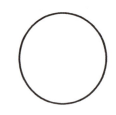

Show the circle partitioned into halves.

How many equal parts is the whole partitioned into?

2

Show the partitioned circle and the answer choices.

Is the whole partitioned into halves, thirds, or fourths?

Halves

Show the answer: Halves

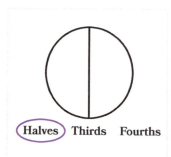

Repeat the process with the following sequence:

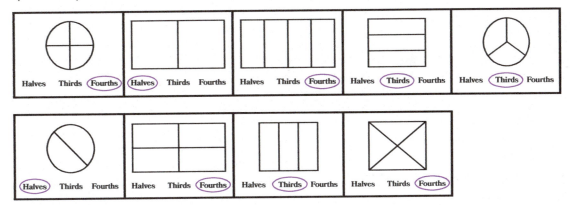

Counting the Math Way by Tens, Fives, and Twos

Students construct a number line with their fingers while counting aloud to build fluency with counting by tens, fives, and twos and develop a strategy for multiplying.

For each skip-count, show the math way on your own fingers while students count, but do not count aloud.

Let's count the math way. Each finger represents 10.

Have students count the math way by tens from 0 to 100 and then back down to 0.

Now let's count the math way by fives. Each finger represents 5.

Have students count the math way by fives from 0 to 50 and then back down to 0.

Teacher Note

Control the pace of the count with your hands. Remember to listen to student responses and be mindful of errors, hesitation, and lack of full-class participation. If needed, adjust the tempo or limit the range of numbers.

Now let's count the math way by twos. Each finger represents 2.

Have students count the math way by twos from 0 to 20 and then back down to 0.

| 0 | 2 | 4 | 6 | 8 | 10 | 12 | 14 | 16 | 18 | 20 |

I Say, You Say: 5 of a Unit

Students say the value of a number given in unit form to prepare for using $5 + n$ with the distributive property beginning in topic C.

Invite students to participate in I Say, You Say.

When I say a number in unit form, you say its value. Ready?

When I say 5 tens, you say 50.

5 tens

50

5 tens

50

When I say 5 fives, you say 25. Ready?

5 fives

25

5 fives

25

Repeat the process with 5 twos.

When students are ready, repeat the sequence, but this time do not provide the value before beginning the chant (i.e., "When I say 5 tens, you say …?").

Teacher Note

The repetition in the vignette is intentional. Consider adding energy to the routine in one or more of the following ways:

- Use the call and response like an upbeat chant heard at sporting events.

- Build to a quick pace.

- Use gestures such as leaning in, pointing, or cupping your ear to signal students to respond.

Launch

Students determine 5 as either the number of groups or the number in each group.

Show pictures of ducks and strawberries one at a time. For each picture, ask:

Is 5 the number of groups or the number in each group?

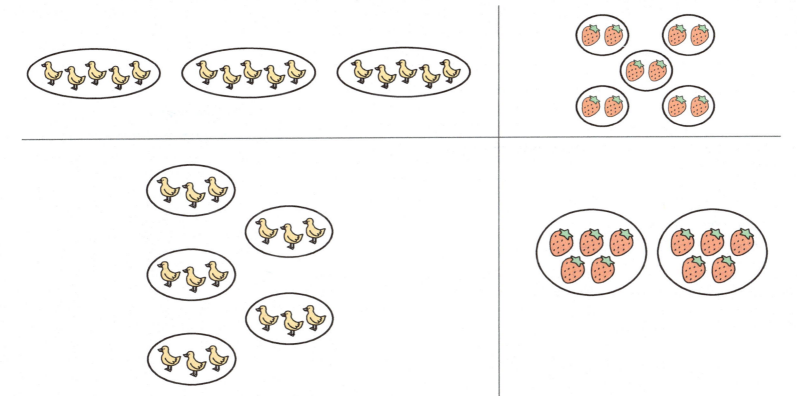

Invite students to turn and talk about how they know whether 5 is the number of groups or the number in each group.

Transition to the next segment by framing the work.

Today, we will represent equal sharing using what we know about the number of groups and the number in each group.

Equally Share Ten Crackers

Materials—S: Paper plates, crackers

Students model, discuss, and compare two interpretations of division.

Give each student pair 5 paper plates and 10 crackers. Direct students to problem 1 in their books and read it chorally. Then ask:

What does it mean to share equally?

To share equally means to give each person the same amount of something.

Invite partners to work together to model problem 1 with their plates and crackers. Then direct students to complete the rest of problem 1. Circulate as partners work and provide support as needed.

UDL: Engagement

To promote relevance of the concept of equal sharing, make connections to contexts that are familiar to students. For example, before asking about what it means to share equally, invite students to think about a time they had to share a snack with friends. Ask students to indicate, by a show of fingers, how many friends shared the snack. Tell students to give a thumbs up if they think their snack was shared equally.

1. Use 10 crackers to make equal shares with 5 crackers in each group.

   a. Draw to show how you equally shared the crackers and then complete the sentences.

   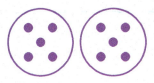

   b. The total number is ____10____.

   c. The number in each group is ____5____.

   d. The number of equal groups is ____2____.

Gather the class to discuss the equal sharing of 10 crackers in groups of 5. Display a completed copy of the problem to support the discussion.

What does 10 represent in this problem? How do you know?

10 represents the total because there are 10 crackers in all.

It's the total because there are 10 dots in my equal groups drawing.

What does 5 represent in this problem? How do you know?

5 is the number in each group because I have 5 crackers on each plate.

It's the number in each group because there are 5 dots in each equal group.

How does knowing that 5 is the number in each group help you equally share the crackers?

We put 5 crackers on a plate and kept doing that until we used all 10 crackers.

Invite students to turn and talk about how they equally shared.

Teacher Note

There are two distinct interpretations of division: *partitive* and *measurement*.

In partitive division, the total and the number of groups are known, but the number in each group is unknown. Partitive division asks the question, "How many are in each group?"

In measurement division, the total and number in each group are known, but the number of groups is unknown. Measurement division asks the question, "How many groups are there?"

Students are not expected to know the terms *partitive division* and *measurement division*, but they are expected to identify which number in a situation or equation represents the number of groups and which represents the number in each group. In later lessons, students describe a division situation in terms of what is known and what is unknown.

Language Support

As students work through the tasks, refer them to the Talking Tool. This tool can help guide their discussions when they aren't sure what to say or how to start.

Invite partners to work together to model problem 2 with their plates and crackers and then complete the rest of problem 2. Circulate as partners work and provide support as needed.

2. Use 10 crackers to make 5 equal groups of crackers.

   a. Draw to show how you equally shared the crackers. Then complete the sentences.

   b. The total number is ___10___.

   c. The number in each group is ___2___.

   d. The number of equal groups is ___5___.

Gather the class to discuss how 10 crackers were equally shared in 5 groups. Display a completed copy of the problem to support the discussion.

What does 10 represent in this problem? How do you know?

10 represents the total because there are 10 crackers in all.

What does 5 represent in this problem? How do you know?

5 is the number of groups because there are 5 plates.

How does knowing that 5 is the number of groups help you equally share the crackers?

We knew we needed 5 groups, so we put 5 plates on the desk. Then we equally shared the crackers until we used all 10 crackers.

Invite students to turn and talk about how they equally shared.

UDL: Representation

Consider creating and posting a chart to clarify the difference between the phrases *number of groups* and *number in each group*. Use numbers that are not included in this lesson to provide an additional example and to help students generalize both phrases.

Number in each group: There are 4 in each group.

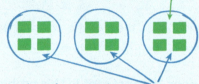

Number of groups: There are 3 groups.

Color coding and using different shapes for each phrase may further support students' understanding.

Promoting the Standards for Mathematical Practice

Students make sense of problems and persevere in solving them **(MP1)**. They determine what each problem asks and attempt to find a solution strategy.

Ask the following questions to promote MP1:

- What are some things you could try to start solving the problem?

- Is your strategy for making equal shares working? What's another way you could try?

Display the two pictures of plates with crackers. Guide students to discuss the similarities and differences between the equal sharing in problems 1 and 2. Fill in the blanks as students discuss.

Problem 1	**Problem 2**

The total number is _____.

The number in each group is _____.

The number of equal groups is _____.

The total number is _____.

The number in each group is _____.

The number of equal groups is _____.

How is the equal sharing in problems 1 and 2 similar? Different?

In both problems, we equally shared a total of 10 crackers.

The number in each group is 5 in problem 1. The number of equal groups is 5 in problem 2.

Invite students to turn and talk about the two different ways they equally shared.

Equally Share Twenty Crackers

Materials—S: Paper plates, crackers

Students model, discuss, and compare two interpretations of division with a larger total.

Give each student pair 10 more crackers. Direct students to problem 3 and read it chorally. Invite partners to work together to model problem 3 with their plates and crackers and then complete the rest of problem 3. Circulate as students work and provide support as needed.

3. Use 20 crackers to make equal shares, with 5 crackers in each group.

   a. Draw to show how you equally shared the crackers, and then complete the sentences.

   b. The total number is ____20____.

   c. The number in each group is ____5____.

   d. The number of equal groups is ____4____.

Gather the class to discuss the equal sharing of 20 crackers in groups of 5. Display a completed solution to the problem to support this discussion.

What does 20 represent in this problem? How do you know?

20 represents the total because there are 20 crackers in all.

What does 5 represent in this problem? How do you know?

5 is the number in each group because I have 5 crackers on each plate.

How does knowing that 5 is the number in each group help you equally share the crackers?

We put 5 crackers on a plate and kept doing that until we used all 20 crackers.

Invite students to turn and talk about how they equally shared.

Direct students to problem 4 and read it chorally. Invite partners to work together to model problem 4 with their plates and crackers and then complete the rest of problem 4. Circulate as students work and provide support as needed.

4. Use 20 crackers to make 5 equal groups of crackers.

    a. Draw to show how you equally shared the crackers. Then complete the sentences.

    b. The total number is ___20___.

    c. The number in each group is ___4___.

    d. The number of equal groups is ___5___.

Gather the class to discuss the equal sharing of 20 crackers in 5 groups. Display a completed copy of the problem to support this discussion.

What does 20 represent in this problem? How do you know?

20 represents the total because the first sentence says use the 20 crackers.

What does 5 represent in this problem? How do you know?

5 is the number of groups because I have 5 plates.

How does knowing that 5 is the number of groups help you equally share the crackers?

We knew we needed 5 groups, so we put 5 plates on the desk. Then we equally shared the crackers until we used all 20 crackers.

Invite students to turn and talk about how they equally shared.

Display the two pictures of plates with crackers. Guide students to discuss the similarities and differences between the equal sharing in problems 3 and 4. Fill in the blanks as students discuss.

Problem 3

The total number is _____.

The number in each group is _____.

The number of equal groups is _____.

Problem 4

The total number is _____.

The number in each group is _____.

The number of equal groups is _____.

How is the equal sharing in problems 3 and 4 similar? Different?

In both problems, we equally shared a total of 20 crackers.

4 is the number of equal groups in problem 3. 4 is the number in each group in problem 4.

How is the equal sharing that you did today different from the work we've done with multiplication? How is it similar?

Today we knew the total and the number in each group or the number of equal groups.

When we multiply, we know the number in each group and the number of equal groups, but we don't know the total.

We can use equal groups for both multiplication and equal sharing but in different ways.

Problem Set

Differentiate the set by selecting problems for students to finish independently within the timeframe. Problems are organized from simple to complex.

Land 10

Debrief 5 min

Objective: Explore measurement and partitive division by modeling concretely and drawing.

Gather students with their completed work. Use the following prompts to guide a discussion about the two interpretations of division. Consider displaying the plates of crackers to support the discussion.

What does the number 5 represent in problems 1 and 3? Problems 2 and 4?

5 represents the number in each group in problems 1 and 3.

5 represents the number of equal groups in problems 2 and 4.

What are the different ways you equally shared 10 crackers? 20 crackers?

We shared 10 crackers by first showing 5 as the number of crackers in each group. Then we used 5 as the number of equal groups.

We did the same thing for 20 crackers. We started by showing 5 as the number of crackers in each group. Then 5 was also the number of equal groups.

Why is it helpful to think about what the numbers in a division problem represent?

It helps me make equal groups. If I know that it's the number in each group, then I can keep making groups with that number until I get to the total.

If it's the number of groups, I make that many groups and then share the total number equally in each group.

Exit Ticket 5 min

Provide up to 5 minutes for students to complete the Exit Ticket. It is possible to gather formative data even if some students do not complete every problem.

Sample Solutions

Expect to see varied solution paths. Accept accurate responses, reasonable explanations, and equivalent answers for all student work.

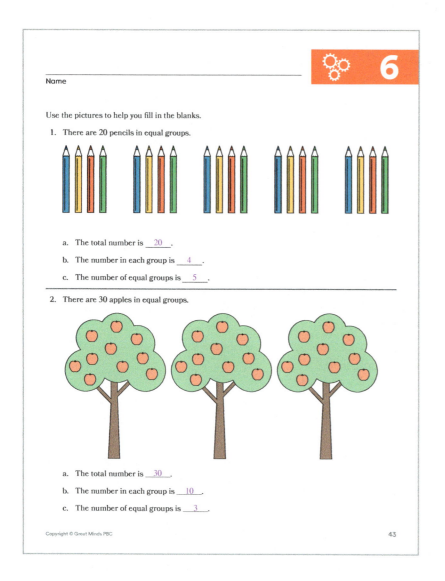

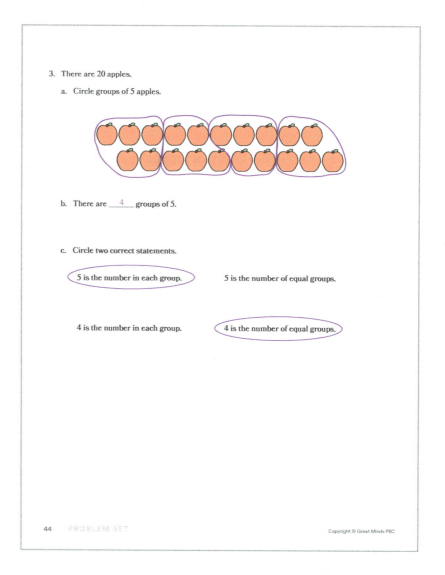

d. Circle the apples to make 5 equal groups.

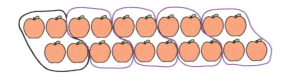

e. There are ___4___ apples in each group.

f. Circle two correct statements.

5 is the number in each group.   ⬭5 is the number of equal groups.⬭

⬭4 is the number in each group.⬭   4 is the number of equal groups.

g. In parts (a) and (d) you equally shared 20 apples.

   How was the equal sharing different?

   Part (a) has 4 groups with 5 apples in each group. Part (d) has 5 groups with 4 apples
   in each group.

Model measurement and partitive division by drawing equal groups.

Name _____

Ray divides 30 beads into 5 equal groups.

a. Draw an equal groups picture to show Ray's beads.

b. How many beads are in each group?

There are 6 beads in each group.

Lesson at a Glance

Students reason about equal groups division situations. Students represent the division situations with drawings and multiplication equations and relate division to multiplication. This lesson formalizes the terms *divide* and *division*.

Key Questions

- How is dividing like multiplying? How is it different?
- Why is it helpful to think about whether we know the number in each group or the number of equal groups when dividing?

Achievement Descriptor

3.Mod1.AD2 **Represent** a division situation with a model and **convert** between several representations of division. (3.OA.A.2)

Agenda

Fluency 10 min

Launch 5 min

Learn 35 min

- Divide Ten Balloons
- Divide Twenty Pieces of Candy
- Problem Set

Land 10 min

Materials

Teacher

- None

Students

- None

Lesson Preparation

None

Fluency ⏱ 10

Choral Response: Equal Parts

Students identify and describe equal parts of a shape to maintain geometry concepts from grade 2.

After asking each question, wait until most students raise their hands, and then signal for students to respond.

Raise your hand when you know the answer to each question. Wait for my signal to say the answer.

Display the rectangle.

What is the name of this shape?

Rectangle

This is 1 whole rectangle.

Show the rectangle partitioned into halves.

How many equal parts is the whole partitioned into?

2

Show the partitioned rectangle and the answer choices.

Is the whole partitioned into halves, thirds, or fourths?

Halves

Show the answer: Halves.

How many halves make the whole?

2 halves

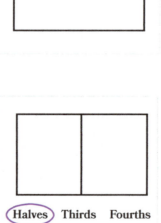

Repeat the process with the following sequence:

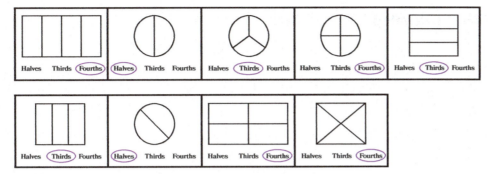

Counting the Math Way by Twos and Tens

Students construct a number line with their fingers while counting aloud to build fluency with counting by tens and twos and develop a strategy for multiplying.

For each skip-count, show the math way on your own fingers while students count, but do not count aloud.

Let's count the math way by twos. Each finger represents 2.

Have students count the math way by twos from 0 to 20 and then back down to 0.

Now let's count the math way by tens. Each finger represents 10.

Have students count the math way by tens from 0 to 100.

Show me 50.

(Students show 50 on their fingers the math way.)

Have students count the math way by tens from 50 to 100 and then back down to 50.

Offer more practice counting the math way by tens, emphasizing counting on from 50.

Teacher Note

In this activity, the students have progressed from starting at 0 to starting with numbers other than 0. In this instance, students start with all five fingers and say "50." Then they count up and count down from there. They know from previous experience that if each finger represents a unit of ten, then holding up five fingers represents 50.

Tap, Tap, Clap Threes

Students count with an emphasis on multiples of three to develop fluency with counting by threes.

Show students the rhythm of tap, tap, clap (i.e., soft tap thighs, soft tap thighs, loud clap hands). For each soft tap of the thighs, tell students they should whisper the number. When they get to a clap, they will call out the number.

Model the action: Clap as you say "0," tap as you whisper "1," tap as you whisper "2," clap as you say "3."

Tap as you whisper "4," tap as you whisper "5," clap as you say "6."

> **Let's try it together.**
>
> (clap) 0
>
> (tap 1), (tap 2), (clap) 3
>
> (tap 4), (tap 5), (clap) 6
>
> (tap 7), (tap 8), (clap) 9

Continue counting to 30, whispering on each tap and saying the multiples of 3 in a louder voice with each clap.

Count backward from 30 to 0, slowing the pace. Whisper on each tap and say the multiples of 3 in a louder voice with each clap.

Launch ⏱ 5

Students determine whether 5 represents the number of equal groups or the number in each group.

Display the equal groups pictures one at a time. For each picture, ask the following:

> **Is 5 the number of equal groups or the number in each group?**

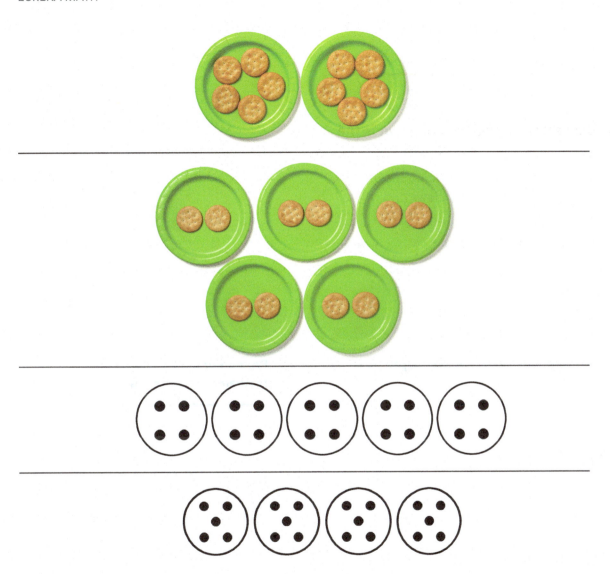

Invite students to turn and talk about how they know whether 5 is the number of equal groups or the number in each group.

Transition to the next segment by framing the work.

Today, we will draw equal groups using what we know about the number of groups and the number in each group.

Divide Ten Balloons

Students watch, discuss, and model a division story, where the unknown is the number of groups.

Gather the class and set the context for the video. Tell students that the video shows a boy preparing for a party. Ask students to watch to figure out what the boy knows and what he is trying to find.

Play part 1 of the Party Problem video, which shows a boy thinking about how many groups of two balloons can be made from ten balloons.

What does the student know? How do you know?

He knows the total because he has 10 balloons.

He knows the number in each group because he puts 2 balloons in a group.

Is he trying to find the number in each group or the number of equal groups? How do you know?

He's trying to find the number of equal groups because he's not sure how many groups of 2 balloons he can make.

Let's draw to find the number of equal groups.

Direct students to the picture of the balloons in their books. Interactively model equal sharing when the number in each group is known. Cross off 2 balloons in the picture and draw a circle with 2 dots in it to represent one group of 2 balloons. Encourage students to continue in this manner until all 10 balloons are equally shared.

Promoting the Standards for Mathematical Practice

Students reason quantitatively and abstractly (MP2) when they create abstract models and equations based on the context of the video.

Ask the following questions to promote MP2:

- What does the video tell you about your equal groups picture?

- Does your answer make sense with what happens in the video?

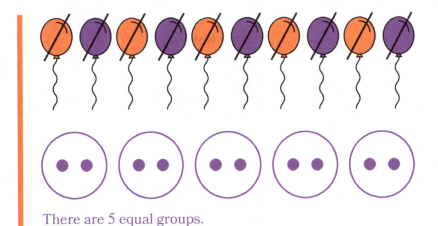

There are 5 equal groups.

Facilitate a class discussion about how the equal groups drawing represents the problem.

Where do you see the total in your equal groups drawing?

I see 10 balloons represented by 10 dots.

Where do you see the number in each group?

The 2 dots in each circle represent the 2 balloons in each group.

Where do you see the number of equal groups?

The 5 circles represent the 5 equal groups of balloons.

If you know that 2 is the number in each group, how does this help you solve the problem?

If I know that 2 is the number in each group, I can keep drawing groups of 2 until I reach the total. Then I can count the number of equal groups in my drawing.

Differentiation: Support

Students may benefit from acting out the balloon division scenario from the video with their partner. Consider providing manipulatives for students to use to support their understanding. Make connections between the concrete manipulatives and the pictorial representations with prompts, such as:

- Show me how you would make groups of 2.

- What can you draw to show the number of twos you made?

- How can you describe what you acted out and what you drew?

Prompt students to write a solution statement.

Let's use a multiplication equation to describe this problem.

Write _____ $\times 2 = 10$.

What does 2 represent?

Label 2 as the number in each group.

What does 10 represent?

Label 10 as the total.

$$\underline{\hspace{2cm}} \times 2 = 10$$

_____ **x** **2** **=** **10**

number of equal groups   number in each group   total

What number should I write in the blank? How do you know?

I should write 5 because I know that $5 \times 2 = 10$.

I should write 5 because my equal groups drawing shows 5 groups with 2 in each group. This makes a total of 10.

Label the blank as the number of equal groups.

The multiplication equation shows that we know the total and the number in each group. The blank represents the number of equal groups, which we don't know.

We just used equal groups and multiplication to help us think about an equal-sharing problem. Another way to think about an equal-sharing problem is to use the word divide. This type of problem is called a division problem.

Display the statement: The boy divides 10 balloons into groups of 2.

Whisper-read the sentence to a partner. What does the word *divide* mean in this sentence?

It means that he equally shares the balloons in groups of 2.

It means that he started with the total and made equal groups.

We just saw that 10 balloons divided into groups of 2 makes 5 groups. Let's finish watching to see if that is what the boy found.

Continue to play the video to show the boy making 5 groups of 2 balloons.

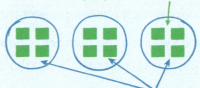

What strategy does the boy use to find the number of equal groups?

He starts with 10 balloons and puts 2 balloons in each group until he runs out of balloons. He makes 5 groups.

How does this match what we did with our drawing?

We drew groups of 2 balloons until we ran out of balloons.

Invite students to turn and talk about how knowing the total and the number in each group helped to find the number of groups.

Divide Twenty Pieces of Candy

Students watch, discuss, and model a division story, where the unknown is the number in each group.

Let's see what else the boy does to get ready for his party.

Play part 2, which shows the boy thinking about how he can equally share 20 candies among 5 bags. Ask students to watch to figure out what the boy knows and what he is trying to find.

What does the boy know? How do you know?

He knows that 20 is the total, and he knows the number of equal groups because there are 5 bags.

What is he trying to find? How do you know?

He's trying to find the number in each group because he's not sure how many pieces of candy to put in each bag.

Why do you think his friends look confused?

He didn't share the candy equally. Some friends have more candy than others.

What can he do differently to divide the candy into equal groups?

He can put one piece of candy in each bag until all the candy is gone.

He can put 4 pieces of candy in each bag.

Let's draw to find the number of candies in each group.

Direct students to the picture of candy in their books. Interactively model equal sharing when the number of groups is known. Draw 5 circles to represent 5 groups. Cross out one piece of candy and draw a dot in the first circle. Cross out another piece of candy and put a dot in the second circle. Continue to equally share in this manner until all 20 pieces of candy are shared.

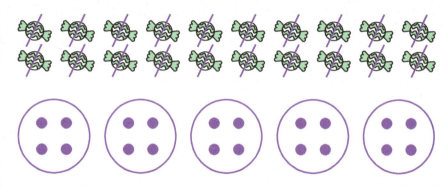

There are 4 pieces of candy in each group.

Facilitate a class discussion about how the equal groups drawing represents the problem.

Where do you see the total in your equal groups drawing?

I see 20 pieces of candy represented as 20 dots.

Where do you see the number of equal groups?

There are 5 circles that represent 5 groups.

Where do you see the number in each group?

There are 4 dots in each circle to represent 4 pieces of candy in each group.

If you know that 5 is the number of groups, how does this help you solve the problem?

If I know that 5 is the number of groups, that helps me because I can draw 5 circles. Then I can equally share until I reach the total. I can count the number of dots in one circle to find the number in each group.

Prompt students to write a solution statement.

Let's use a multiplication equation to describe this problem.

Write $5 \times$ ____ $= 20$.

What does the 5 represent?

Label 5 as the number of equal groups.

What does the 20 represent?

Label 20 as the total.

$$5 \quad \times \quad \underline{\hspace{2cm}} \quad = \quad 20$$

number of equal groups   number in each group   total

What does the blank represent?

Label the blank as the number in each group.

What number should I write in the blank? How do you know?

You should write 4 because we know that $5 \times 4 = 20$.

You should write 4 because the equal groups drawing shows 5 groups and 4 in each group. This makes a total of 20.

Let's watch to see how the boy divides the candy into equal groups.

Play part 3, which shows the boy equally sharing 20 candies among 5 bags.

How does the boy use what he knows—the number of equal groups—to divide the candy?

He knows he needs to make 5 equal groups, so he puts a piece of candy in each bag until he uses all 20 pieces of candy.

Invite students to think–pair–share about the differences between what was known and unknown when he equally shared the balloons and the candies.

When he divided the balloons, he put 2 balloons in each group, but he wasn't sure how many groups he could make.

When he shared the candy, he knew he had 5 bags, but he didn't know how many pieces of candy to put in each bag.

Invite students to turn and talk about how the equal groups drawings represent the problems.

As time allows, continue the same process with the following problems:

- The boy puts 12 blueberries on some cupcakes. He puts 2 blueberries on each cupcake. How many cupcakes get blueberries?
- The boy has 30 pieces of gum. He wants to divide the gum equally among the 5 bags. How many pieces of gum can he put in each bag?

Display the following sentences.

The boy divides 10 balloons into groups of 2.	The boy divides 20 pieces of candy into 5 groups.
The boy divides 12 blueberries into groups of 2.	The boy divides 30 pieces of gum into 5 groups.

These sentences describe the different ways the boy divides to get ready for his party. What is the total in each sentence?

Circle the total in each sentence.

Why do we need to know the total to divide?

To equally share something, first you need to know how many you have. Dividing is the same as equally sharing.

Problem Set

Differentiate the set by selecting problems for students to finish independently within the timeframe. Problems are organized from simple to complex.

Land 10

Debrief 5 min

Objective: Model measurement and partitive division by drawing equal groups.

Initiate a class discussion using the prompts below. To support this discussion, invite students to refer to their completed problems.

What are the two different ways we divided today?

We divided 10 into equal groups of 2. We knew the total and the number in each group. We didn't know the number of equal groups.

We divided 20 into 5 equal groups. We knew the total and the number of groups. We didn't know the number in each group.

How is dividing like multiplying? How is it different?

They both use number of equal groups, number in each group, and a total.

When we multiply, we start with the number of groups and the number in each group. Then we find the total.

When we divide, we start with the total and the number in each group or the number of groups. Then we find the number of groups or the number in each group.

Why is it helpful to think about whether we know the number in each group or the number of equal groups when dividing?

If we know the number in each group, then we know we're trying to find the number of equal groups.

If we know the number of equal groups, then we know we're trying to find the number in each group.

It makes it easier to draw an equal groups picture. When we know the number in each group, we can keep drawing equal groups of that number until we get to the total.

When we know the number of groups, we can draw that many groups and then equally share until we get to the total.

Exit Ticket 5 min

Provide up to 5 minutes for students to complete the Exit Ticket. It is possible to gather formative data even if some students do not complete every problem.

Sample Solutions

Expect to see varied solution paths. Accept accurate responses, reasonable explanations, and equivalent answers for all student work.

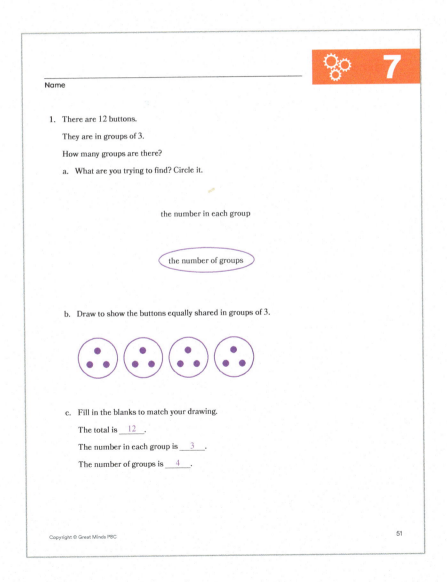

Name _____

7

1. There are 12 buttons.

   They are in groups of 3.

   How many groups are there?

   a. What are you trying to find? Circle it.

      the number in each group

      ⟨the number of groups⟩

   b. Draw to show the buttons equally shared in groups of 3.

   c. Fill in the blanks to match your drawing.

      The total is   12   .

      The number in each group is   3   .

      The number of groups is   4   .

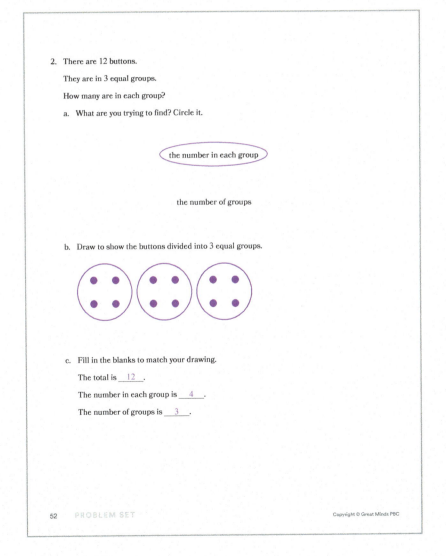

2. There are 12 buttons.

   They are in 3 equal groups.

   How many are in each group?

   a. What are you trying to find? Circle it.

      ⟨the number in each group⟩

      the number of groups

   b. Draw to show the buttons divided into 3 equal groups.

   c. Fill in the blanks to match your drawing.

      The total is   12   .

      The number in each group is   4   .

      The number of groups is   3   .

3. There are 30 crackers.

   a. Draw to show the crackers divided into 5 equal groups.

   b. How many crackers are in each group?

   There are 6 crackers in each group.

   c. Draw to show the crackers divided into groups of 5.

   d. How many groups of crackers are there?

   There are 6 groups of crackers.

   e. Which drawing shows 5 as the number in each group? How do you know?

   The second drawing shows 5 as the number in each group. There are 5 crackers in each
   group because 30 crackers were divided equally into 6 groups.
   5 + 5 + 5 + 5 + 5 + 5 = 30 or 6 × 5 = 30

4. Adam draws a picture to show 15 divided into 5 equal groups.

   Do you agree with Adam's work? Explain.

   I do not agree with Adam's work. He divided 15 into 3 equal groups with 5 in each group.
   He should have divided 15 into 5 equal groups, so there would be 5 groups with 3 dots
   in each group.

Copyright © Great Minds PBC

PROBLEM SET   53

54   PROBLEM SET

Copyright © Great Minds PBC

126

Copyright © Great Minds PBC

8

Model measurement and partitive division by drawing arrays.

Name _____ ✉ **8**

Ray puts 18 cards into rows of 3.

a. Draw an array to find the number of rows.

○ ○ ○
○ ○ ○
○ ○ ○
○ ○ ○
○ ○ ○
○ ○ ○

b. Write a division equation to show the number of rows.

$18 \div 3 = 6$

c. How many rows does Ray make?

___6___ **rows**

65

Lesson at a Glance

Students relate equal groups to arrays and write division equations to represent division situations. This lesson introduces the division symbol, ÷, and the division equation.

Key Questions

- How can arrays be used to represent division?
- How can equations be used to represent division?

Achievement Descriptors

3.Mod1.AD2 **Represent** a division situation with a model and **convert** between several representations of division. (3.OA.A.2)

3.Mod1.AD3 **Solve** one-step word problems by using multiplication and division within 100, involving factors and divisors 2–5 and 10. (3.OA.A.3)

Agenda

Fluency 10 min

Launch 5 min

Learn 35 min

- Arrays and Equations to Represent Division
- Same Equation, Different Situation
- Problem Set

Land 10 min

Materials

Teacher

- Interlocking cubes, 1 cm (20)
- Straightedge

Students

- Straightedge

Lesson Preparation

Prepare 20 interlocking cubes in one color.

Fluency ⑩

Counting the Math Way by Twos and Fives

Students construct a number line with their fingers while counting aloud to build fluency with counting by twos and fives and develop a strategy for multiplying.

For each skip-count, show the math way on your own fingers while students count, but do not count aloud.

> **Let's count the math way by twos. Each finger represents 2.**

Have students count the math way by twos from 0 to 20 and then back down to 0.

> **Now let's count the math way by fives. Each finger represents 5.**

Have students count the math way by fives from 0 to 50.

> **Show me 25.**
>
> (Students show 25 on their fingers the math way.)

Have students count the math way by fives from 25 to 50 and then back down to 25.

Offer more practice counting the math way by fives emphasizing counting on from 25.

Tap, Tap, Clap Threes

Students count with an emphasis on multiples of three to develop fluency with counting by threes.

Show students the rhythm of tap, tap, clap (i.e., soft tap thighs, soft tap thighs, loud clap hands). For each soft tap of the thighs, tell students to whisper the number. When they get to a clap, they call out the number.

Model the action: Clap as you say "0," tap as you whisper "1," tap as you whisper "2," clap as you say "3."

Tap as you whisper "4," tap as you whisper "5," clap as you say "6."

Let's try it together.

(clap) 0

(tap 1), (tap 2), (clap) 3

(tap 4), (tap 5), (clap) 6

(tap 7), (tap 8), (clap) 9

Continue counting to 30, whispering on each tap and saying the multiples of 3 in a louder voice with each clap.

Count backward from 30 to 0, slowing the pace. Whisper on each tap, and say the multiples of 3 in a louder voice with each clap.

Count from 0 to 30 again, but this time encourage students to think of the number in their head on each tap and say only the multiples of 3 aloud.

Choral Response: Relating Multiplication Models

Students relate an equal groups picture, array, or tape diagram with a unit of 2 or 3 to a repeated addition expression, unit form, and multiplication equation to build an understanding of multiplication.

After asking each question, wait until most students raise their hands, and then signal for students to respond.

Raise your hand when you know the answer to each question. Wait for my signal to say the answer.

Display the picture of 2 groups of 3 apples.

What repeated addition expression represents this picture?

$3 + 3$

How do you represent the picture in unit form?

2 threes

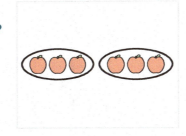

What multiplication equation represents the picture?

$2 \times 3 = 6$

Repeat the process with the following sequence:

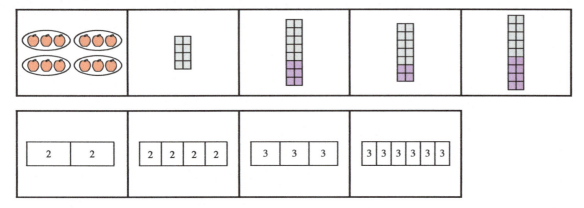

Launch ⏱ 5

Students discuss the similarities and differences between equal groups and arrays.

Display the pictures of the eggs arranged in groups and an array.

Picture A

Picture B

Invite students to think–pair–share about how pictures A and B are similar and how they are different. Circulate as partners discuss, and use the following questions to advance their thinking:

- How are the eggs arranged in picture A? In picture B?
- What is the total number in picture A? In picture B?
- Where do you see 5 in picture A? In picture B?
- Where do you see 4 in picture A? In picture B?

Gather the class and invite students to share what they discussed with their partner. After giving students an opportunity to share, ask:

Do you think these pictures represent multiplication or division?

The equal groups picture could represent either. We use equal groups to represent multiplication and division.

I think the equal groups picture could represent either, but I'm not sure about the array. I wonder if we can use arrays to show division.

Transition to the next segment by framing the work.

We just used both arrays and equal groups to represent multiplication. But we've only used equal groups to represent division. Today, we will use arrays to divide.

Arrays and Equations to Represent Division

Materials—T: Cubes, straightedge; S: Straightedge

Students draw arrays and write equations to represent measurement and partitive division.

Direct students to problem 1 and read it chorally. Ask the class what they know and what they are trying to find.

Use the Read–Draw–Write process to solve the problem.

1. Gabe collects seashells.

   He has 20 seashells that he wants to put in rows of 5.

   How many rows of 5 seashells can Gabe make?

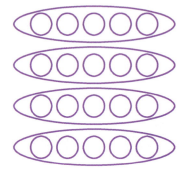

Gabe can make 4 rows of 5 seashells.

$20 \div 5 = 4$

Teacher Note

The Read–Draw–Write process used in problems 1 and 2 is slightly different from the traditional RDW process. The teacher uses concrete materials (i.e., cubes) to represent the problem instead of drawing, and the solution statement is written before the equation. This change in process allows for formal introduction of the division equation using the newly learned division symbol.

Display a pile of 20 loose cubes on a whiteboard.

Each cube represents a seashell. Watch as I use the cubes to help us find how many rows Gabe can make with his seashells. I am going to start with the total, 20. Then I am going to make rows of 5.

Model making a row of 5 with the interlocking cubes.

Here's one row of 5.

Invite students to draw a row of 5 small circles to represent the cubes.

Repeat the process of teacher modeling while students draw until you have used all 20 cubes.

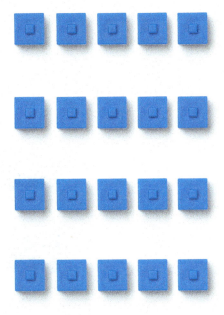

What is the name of the model we just made?

It's an array.

How can we use the array to find how many rows of seashells Gabe can make?

We can count the number of rows in the array.

Let's circle each row as we count.

Model circling and counting each row of the cube array as students do the same.

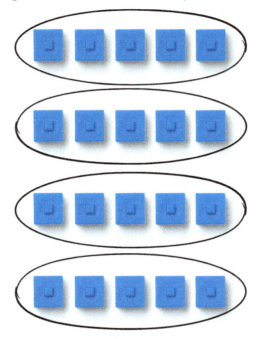

Differentiation: Support

To support students in seeing 4 fives, consider linking the cubes in each row together instead of circling.

- How many fives are in 20?

- Let's count. 1 five, 2 fives, 3 fives, 4 fives.

- There are 4 fives in 20.

How many rows of 5 seashells can Gabe make?

Direct students to write a solution statement.

We started with the total, 20, and made rows of 5 until all 20 cubes were used. Did we multiply or divide? How do you know?

We divided because we started with the total and shared it in rows of 5 equally.

We can represent how we divided 20 into rows of 5 by writing a division equation.

Think aloud as you write $20 \div 5 = 4$.

We started with the total, 20. We divided 20 into rows of 5. We figured out that there are 4 rows.

We use the symbol ÷ to show division.

Write *divided by* and draw an arrow from the words *divided by* to the symbol. Support student understanding of the division equation by labeling each number as shown.

Have students copy the division equation.

To read the division equation, we say *twenty divided by five equals four*. Whisper read the division equation to a partner.

divided by

$$20 \div 5 = 4$$

total   number   number
in each   of rows
row

Guide students to make connections between the division equation, the array, and the context of the problem. Ask them how each number in the equation relates to the array and the word problem.

Direct students to problem 2 and read it chorally. Ask the class what they know and what they are trying to find.

> Use the Read–Draw–Write process to solve the problem.
>
> 2. Gabe also collects rocks.
>
>   He has 15 rocks that he wants to display in 5 equal rows.
>
>   How many rocks can Gabe put in each row?
>
>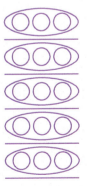
>
>   Gabe can put 3 rocks in each row.
>
> $15 \div 5 = 3$

Differentiation: Support

Provide access to manipulatives, such as interlocking cubes, square tiles, or two-color counters, for direct modeling. This supports students as they transition from concrete to pictorial representations and allows for flexibility in demonstrating learning.

Display a pile of 15 loose cubes on a whiteboard.

Watch as I use the cubes to help us find how many rocks Gabe can put in each row. I'll draw 5 lines to represent the 5 rows.

Invite students to draw 5 lines to represent the 5 rows of rocks.

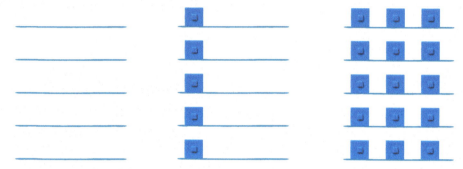

I am going to start with the total, 15, and put one cube in each row.

Model putting one cube in each of the 5 rows. Invite students to draw circles on the 5 lines already drawn to represent the cubes. Repeat the process of teacher modeling while students draw until you have used all 15 cubes.

How can we use the array to find how many rocks Gabe can put in each row?

We can count the number in each row.

Model circling to show the number in each row of the cube array as students do the same.

How many rocks can Gabe put in each row?

Direct students to write a solution statement.

We started with the total, 15, and made 5 equal rows. Did we multiply or divide? How do you know?

We divided because we started with the total and shared it equally in 5 rows.

Let's write a division equation to represent how we divided 15 into 5 rows.

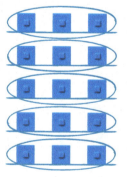

Think aloud as you write $15 \div 5 = 3$.

We started with the total, 15. We divided 15 equally into 5 rows. We figured out that the number in each row is 3.

Support student understanding by labeling each number in the division equation. Have students copy the division equation and whisper-read the division equation to a partner.

$15 \div 5 = 3$

total · number of rows · number in each row

Guide students to make connections between the division equation, the array, and the context of the problem by asking how each number in the equation relates to the array and to the word problem.

Display the pictures of the arrays of cubes representing Gabe's seashells and rocks. Invite students to think–pair–share about how the arrays also show equal groups. Ask partners to share their ideas with the class.

Our arrays look like equal groups that have been neatly lined up.

It's easy to see equal groups in these arrays because we circled the rows. We can see the number of rows and the number in each row.

Look at the number of rows and the number in each row to help you think about arrays as equal groups.

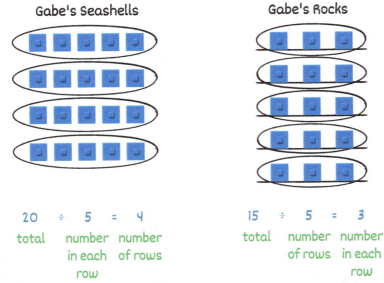

Gabe's Seashells

Gabe's Rocks

$20 \div 5 = 4$

total · number in each row · number of rows

$15 \div 5 = 3$

total · number of rows · number in each row

I noticed that we used the number 5 in both of our division equations. Does 5 represent the same thing in both equations? How do you know?

No, in the division equation for Gabe's seashells, 5 represents the number in each row. In the division equation for his rocks, 5 represents the number of rows.

How does knowing that 5 is the number in each row help you draw an array?

When I know 5 is the number in each row, I keep drawing rows of 5 until I get to the total.

How does knowing that 5 is the number of rows help you draw an array?

When I know 5 is the number of rows, I can draw 5 rows. Then I can draw one circle in each row again and again until I get to the total.

Same Equation, Different Situation

Students draw arrays and write division equations to represent measurement and partitive division word problems.

Direct students to problem 3 and read it chorally. Ask the class what they know and what they are trying to find. Guide students to make a connection between the shelves in the word problems and rows in an array.

Invite students to work with a partner to solve problem 3. Circulate as students work and use the following questions to advance student thinking.

- What does 2 represent in the problem? How does knowing what 2 represents help you draw an array?
- How does the array help you solve the problem?
- What division equation can you write to represent the problem?

Promoting the Standards for Mathematical Practice

Students make sense of problems and persevere (MP1) as they solve word problems and try to find the strategies that work.

Ask the following questions to promote MP1:

- What are some things you could try to start solving the problem?
- Does your drawing make sense with the problem?

Use the Read–Draw–Write process to solve each problem.

3. Jayla stocks shelves at the grocery store.

   a. Jayla has 10 rolls of paper towels. She puts 2 rolls of paper towels on each shelf. How many shelves does Jayla put paper towels on?

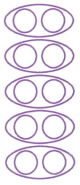

$10 \div 2 = 5$

Jayla puts paper towels on 5 shelves.

b. Jayla has 10 cereal boxes. She puts an equal number of cereal boxes on 2 shelves. How many cereal boxes does Jayla put on each shelf?

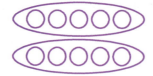

$10 \div 2 = 5$

Jayla puts 5 cereal boxes on each shelf.

Facilitate a discussion about problem 3 by gathering the class. Show the paper towels and cereal boxes array pictures.

What do you notice about the equations in problem 3?

Invite students to turn and talk about how the same equation can represent different division scenarios. Circulate as partners discuss and listen for students to articulate the following:

- The number 10 represents the total in both equations.

- In the paper towels array, 2 represents the number in each row. In the cereal boxes array, 2 represents the number of rows.

- In the paper towels array, 5 represents the number of rows. In the cereal boxes array, 5 represents the number in each row.

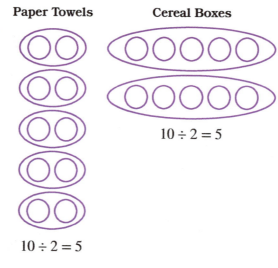

Paper Towels   **Cereal Boxes**

$10 \div 2 = 5$

$10 \div 2 = 5$

Teacher Note

To support the discussion, consider displaying a sample of student work for problem 3 instead of the picture provided.

Problem Set

Differentiate the set by selecting problems for students to finish independently within the timeframe. Problems are organized from simple to complex.

Debrief 5 min

Objective: Model measurement and partitive division by drawing arrays.

Picture A

Picture B

Show the pictures of the eggs arranged in groups and an array.

Gather the class and facilitate a discussion about arrays and division.

How can arrays be used to represent division?

Arrays show the total, the number of rows, and the number in each row.

What are the two ways you used arrays to represent division today?

When we knew the total and the number in each row, we were trying to find the number of rows.

When we knew the total and the number of rows, we were trying to find the number in each row.

How can equations be used to represent division?

We start with the total and divide the total by the number of rows or the number in each row.

Exit Ticket 5 min

Provide up to 5 minutes for students to complete the Exit Ticket. It is possible to gather formative data even if some students do not complete every problem.

Sample Solutions

Expect to see varied solution paths. Accept accurate responses, reasonable explanations, and equivalent answers for all student work.

8

Name _____

Fill in the blanks to match the arrays.

1. There are 8 pencils in equal rows.

   a. The number in each row is ___2___.

   b. The number of rows is ___4___.

   c. ___8___ ÷ ___2___ = ___4___
      total   number in each row   number of rows

2. There are 12 stars in equal rows.

   a. The number in each row is ___4___.

   b. The number of rows is ___3___.

   c. ___12___ ÷ ___4___ = ___3___
      total   number in each row   number of rows

3. Adam puts 14 books on shelves.

   He puts 7 books on each shelf.

   How many shelves does Adam put books on?

   a. Draw an array to represent the problem.

   b. Write a division equation to represent the problem.

   $14 \div 7 = 2$

   c. Adam puts books on ___2___ shelves.

4. There are 21 students in a class.

The students sit in 3 equal rows.

How many students are in each row?

a. Draw an array to represent the problem.

b. Write a division equation to represent the problem.

$21 ÷ 3 = 7$

c. There are ___7___ students in each row.

5. Amy has 9 chairs.

She arranges the chairs in 3 equal rows.

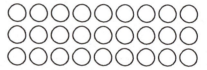

What mistake did Amy make? How do you know?

Amy's mistake is that she drew 9 chairs in each row, but there are only 9 chairs total. She should have drawn 3 chairs in each row.

Represent and solve division word problems using drawings and equations.

 9

Name

Use the Read–Draw–Write process to solve the problem.

Luke has 20 dog treats.

He divides the treats equally among 5 dogs.

How many treats does each dog get?

Each dog gets 4 treats.

71

Lesson at a Glance

Students select strategies to represent and solve division word problems using drawings and equations. After working independently to solve the problems, students share their work to compare and connect the representations and strategies.

Key Questions

- How does thinking about what is known and unknown help you solve division word problems?
- Why is it helpful to compare different strategies that are used to solve the same problem?

Achievement Descriptors

3.Mod1.AD2 **Represent** a division situation with a model and **convert** between several representations of division. (3.OA.A.2)

3.Mod1.AD3 **Solve** one-step word problems by using multiplication and division within 100, involving factors and divisors 2–5 and 10. (3.OA.A.3)

Agenda

Fluency 10 min

Launch 10 min

Learn 30 min

- Number in Each Group Unknown
- Number in Each Group Unknown: Share, Compare, and Connect
- Number of Groups Unknown
- Number of Groups Unknown: Share, Compare, and Connect
- Problem Set

Land 10 min

Materials

Teacher

- Prepared signs

Students

- None

Lesson Preparation

Prepare two signs—one that says *Equal Groups* and one that says *Array*. Hang the signs up in two different locations in the classroom.

Fluency

Counting the Math Way by Twos and Threes

Students construct a number line with their fingers while counting aloud to build fluency with counting by twos, develop fluency with counting by threes, and develop a strategy for multiplying.

For each skip-count, show the math way on your own fingers while students count, but do not count aloud.

Let's count the math way by twos. Each finger represents 2.

Have students count the math way by twos from 0 to 20 and then back down to 0.

Show me 10.

(Students show 10 on their fingers the math way.)

Have students count the math way by twos from 10 to 20 and then back down to 10.

Offer more practice counting the math way by twos, emphasizing counting on from 10.

Now let's count the math way by threes. Each finger represents 3.

Have students count the math way by threes from 0 to 15 as shown and then back down to 0.

| 0 | 3 | 6 | 9 | 12 | 15 |

Choral Response: Relating Multiplication Models

Students relate an array or tape diagram with a unit of 2 or 3 to a repeated addition expression, unit form, and multiplication equation to build an understanding of multiplication.

After asking each question, wait until most students raise their hand, and then signal for students to respond.

Raise your hand when you know the answer to each question. Wait for my signal to say the answer.

Display the 3 by 2 rectangle.

What repeated addition expression represents this picture?

$2 + 2 + 2$

How do you represent the picture in unit form?

3 twos

What multiplication equation represents the picture?

$3 \times 2 = 6$

Repeat the process with the following sequence:

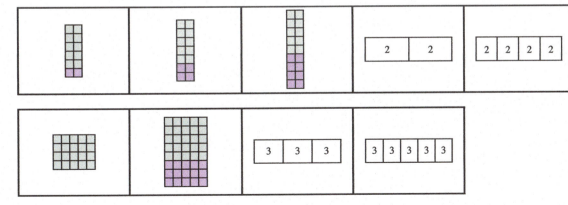

Teacher Note

If students need more repetition using pictures of equal groups, use the pictures provided in lessons 4, 5, and 8.

I Say, You Say: 5 or 2 of a Unit

Students say the value of a number given in unit form to prepare for using $5 + n$ with the distributive property beginning in topic C.

Invite students to participate in I Say, You Say.

When I say a number in unit form, you say its value. Ready?

When I say 5 tens, you say?

50

5 tens

50

5 tens

50

Repeat the process with the following sequence:

5 fives	5 twos	5 threes	2 tens	2 fives	2 twos	2 threes

Teacher Note

Consider using this fluency to reenergize the class throughout the school day. It can be a review of previous learning or used with the current content. Variations could include:

- a different number of each unit (e.g., I say 3 tens; you say 30.);

- measurement conversions (e.g., I say 100 centimeters; you say 1 meter.).

Launch 10

Materials—T: Signs

Students identify and justify their choice of models to represent division word problems.

Introduce the Take a Stand routine to the class. Draw students' attention to the signs hanging in the classroom that say *Equal Groups* and *Array* (see Lesson Preparation).

Display the following problem:

> There are 30 students in the library. The students are working in groups of 3 to complete a project. How many groups of students are working in the library?

Chorally read the problem and ask students to think about which model they would draw to represent the problem—equal groups or an array. Invite students to stand beside the sign that best describes their thinking.

When all students are standing near a sign, allow a few minutes for students to discuss with a partner the reasons why they chose that sign.

Then call on each group to share reasons for their selection. Invite students who change their minds during the discussion to join a different group.

Repeat the process with the following problem:

> The librarian puts 32 books on a table for the students to use. He arranges the books in 4 rows. How many books does the librarian put in each row?

Guide students in understanding that both equal groups and arrays can be used to represent the problems. Have students return to their seats. As a class, reflect on how students decide what to draw to represent a word problem.

Transition to the next segment by framing the work.

Today, we will use our understanding of different division models to solve word problems.

Differentiation: Support

Some students may need to work through the problem to select a preferred representation and explain their choice rather than only processing mentally and verbally. Consider supporting these students by encouraging them to use a whiteboard to show their thinking.

Also consider displaying pictures of the representations instead of just their names to assist students who need visual cues.

Equal Groups   Array

Number in Each Group Unknown

Students reason about, represent, and solve an *equal groups with group size unknown* word problem.

Direct students to problem 1 and chorally read the problem with the class. Have students work independently to use the Read–Draw–Write process to solve the problem. Provide materials such as interlocking cubes for student use. Encourage students to self-select their tools and strategies.

> Use the Read–Draw–Write process to solve the problem.
>
> 1. There are 24 desks in Miss Wong's classroom.
>
>   She arranges the desks into 6 equal groups.
>
>   How many desks does Miss Wong put in each group?
>
> $24 \div 6 = 4$
>
>   Miss Wong puts 4 desks in each group.

Circulate and observe student strategies. Select two or three students to share their strategies in the next segment. Look for work samples that help advance the lesson's objective of using different division models, such as equal groups and arrays.

The student work samples demonstrate using equal groups in different ways to divide.

Equal Groups

$24 \div 6 = 4$

Miss Wong puts 4 desks in each group.

Equal Groups

$24 \div 6 = 4$

Miss Wong puts 4 desks in each group.

Promoting the Standards of Mathematical Practice

Students use appropriate tools strategically (MP5) when they select their own solution strategies and decide which type of model to draw.

Ask the following questions to promote MP5:

- What kind of drawing would be helpful?

- Why did you choose to draw an array with 6 rows?

Differentiation: Support

Consider providing prompts to support students in picturing the scenario and drawing a representation. Circulate and monitor them while they work, and ask the following:

Visualization:

- Close your eyes while I reread the problem. Try to imagine what is happening. What did you see? What can you draw to represent what you saw?

Representation:

- Tell me about your representation of the problem. Why did you choose to use this representation? Is it helpful? How?

- Where is the 24 represented? Is 24 the total number of desks, the number of groups, or the size of each group? How do you know?

- Where is the 6 represented? Is 6 the total number of desks, the number of groups, or the size of each group? How do you know?

Number in Each Group Unknown: Share, Compare, and Connect

Students share solutions for problem 1 and reason about their connections.

Gather the class and invite the students you identified in the previous segment to share their solutions one at a time. As each student shares, ask questions to elicit their thinking and clarify the model used to represent the problem. Ask the class questions to make connections between the different solutions and their own work. Encourage students to ask questions of their own.

Equal Groups (Carla's Way)

Carla, tell us about your drawing.

I drew squares to represent the 24 desks. Then I made 6 equal groups by circling 4 desks in each group. I saw that 1, 2, or 3 desks in each group were not enough.

What did we have to find, or was unknown, in this problem?

We had to find the number in each group.

How does drawing what is known help solve for the unknown?

My drawing shows 6 equal groups with 4 desks in each group, so that's how I know that the number in each group is 4.

What division equation represents the problem? Why?

$24 \div 6 = 4$ because 24 is the total and we divided that by 6, the number of groups. We found that 4 is the number in each group.

Invite students to turn and talk about the similarities and differences between Carla's work and their work.

$24 \div 6 = 4$

Miss Wong puts 4 desks in each group.

Language Support

Consider having the Talking Tool available to assist students when asking and answering questions.

Equal Groups (Gabe's Way)

Gabe, tell us about your drawing.

My drawing shows 6 circles to represent the 6 equal groups. After I drew the circles, I drew a dot in each circle until I got to 24 dots.

$$24 \div 6 = 4$$

Miss Wong puts 4 desks in each group.

What is unknown in this problem? How does drawing what is known help solve for the unknown?

The unknown is the number in each group. The picture shows 6 groups with 4 dots in each group. Each dot represents a desk, so that's how we know that the number of desks in each group is 4.

How does the equation represent the problem?

24 represents the total, 6 represents the number of groups, and 4 represents the number in each group.

How is Gabe's work similar to Carla's work? How is it different?

They both used equal groups and have the same division equation.

Carla drew 24 squares and then circled the squares to make 6 equal groups. Gabe drew 6 circles and then drew a dot in each circle until he had a total of 24 dots.

Invite students to turn and talk about the similarities and differences between Gabe's work and their work.

Number of Groups Unknown

Students reason about, represent, and solve an *array with number of groups unknown* word problem.

Direct students to problem 2 and chorally read the problem with the class. Have students work independently to use the Read–Draw–Write process to solve the problem. Provide materials such as interlocking cubes for student use. Encourage students to self-select their tools and strategies.

> Use the Read–Draw–Write process to solve the problem.
>
> 2. Miss Wong displays the 24 pictures her students made in art class on a bulletin board.
>
>   She puts 8 pictures in each row.
>
>   How many rows of pictures does Miss Wong make?
>
> $24 \div 8 = 3$
>
>   Miss Wong makes 3 rows of pictures.

Circulate and observe student strategies. Select two or three students to share their strategies in the next segment. Look for work samples that help advance the lesson's objective of using different division models, such as equal groups and arrays.

The student work samples demonstrate using equal groups and an array to divide.

Differentiation: Support

Consider providing students tools such as sticky notes, interlocking cubes, or inch tiles to support in acting out the word problem.

Equal Groups	Array

$24 \div 8 = 3$	$24 \div 8 = 3$
Miss Wong makes 3 rows of pictures.	Miss Wong makes 3 rows of pictures.

Number of Groups Unknown: Share, Compare, and Connect

Students share solutions for problem 2 and reason about their connections.

Gather the class and invite the students you identified in the previous segment to share their solutions one at a time. Consider intentionally ordering shared student work from a representational model (equal groups) to a more abstract model (array).

As each student shares, ask questions to elicit their thinking and clarify the model used to represent the problem. Ask the class questions to make connections between the different solutions and their own work. Encourage students to ask questions of their own.

Equal Groups (Shen's Way)

Shen, tell us about your drawing.

I drew groups of 8 dots. I kept drawing groups of 8 until I got to 24. I circled each group of 8 dots to show the number of groups.

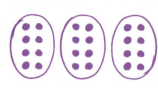

$24 \div 8 = 3$

Miss Wong makes 3 rows of pictures.

What was unknown in this problem? How does drawing what is known help solve for the unknown?

The unknown is the number of groups. The drawing shows a total of 24 in 3 groups of 8, so that's how we know that Miss Wong can make 3 rows of pictures.

How does the equation represent the problem?

24 represents the total, 8 represents the number in each group, and 3 represents the number of groups.

Invite students to turn and talk about the similarities and differences between Shen's work and their work.

Teacher Note

To further investigate student thinking and increase student engagement, consider other questions such as:

- Who can explain Shen's solution strategy?

- Who can restate how Liz figured out the problem in your own words?

- What relationship do you see between Shen's drawing and Liz's drawing? Why did they both draw dots and circle them?

- Who has a similar way to represent and solve the problem? Who has a different way?

Array (Liz's Way)

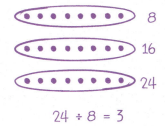

$$24 \div 8 = 3$$

Miss Wong makes 3 rows of pictures.

Liz, tell us about your drawing.

I drew 8 dots in each row until I got to the total, 24. I wrote the total number of dots at the end of each row so I could easily keep track of how many dots I drew. Then I circled each row of 8 dots to show the number of rows.

What was unknown in this problem? How does drawing what is known help solve for the unknown?

The unknown is the number of rows. In the array, there are 3 rows of 8, so we know that Miss Wong makes 3 rows of pictures.

How does the equation represent the problem?

24 represents the total, 8 represents the number in each row, and 3 represents the number of rows.

How is Liz's work similar to Shen's work? How is it different?

They have the same division equation and the same answer.

Shen drew an equal groups picture and Liz drew an array. Shen's drawing shows 3 groups of 8 and Liz's drawing shows 3 rows of 8. The groups and the rows show the same thing.

Invite students to turn and talk about the similarities and differences between Liz's work and their work.

Problem Set

Differentiate the set by selecting problems for students to finish independently within the timeframe. Problems are organized from simple to complex.

UDL: Action & Expression

To support students in monitoring their own progress, consider providing questions that guide self-monitoring and reflection. For example, post the following for students to refer to as they work independently:

- How is the problem like other problems?

- What ways of solving have I or my classmates used to solve problems like this before?

Debrief 5 min

Objective: Represent and solve division word problems using drawings and equations.

Gather students and use the following prompts to guide a discussion about solving division word problems. To support this discussion, display the selected student work for problem 2.

How does thinking about what is known and unknown help you solve division word problems?

Thinking about what is known helps me know if I should draw the number of groups or the number in each group.

If the unknown is the number of groups, I can count the number of groups in my drawing to solve the problem.

If the unknown is the number in each group, I can look at the number in each group in my drawing to solve the problem.

Why is it helpful to compare different strategies that are used to solve the same problem?

It helps me think about the problem in a different way. Maybe I drew equal groups, but someone else drew an array. We can talk about how our drawings show the same things but in different ways.

I can think about how I solved the problem differently from someone else, but we got the same answer. It helps me see that there are different ways to solve the same problem.

Exit Ticket 5 min

Provide up to 5 minutes for students to complete the Exit Ticket. It is possible to gather formative data even if some students do not complete every problem.

Sample Solutions

Expect to see varied solution paths. Accept accurate responses, reasonable explanations, and equivalent answers for all student work.

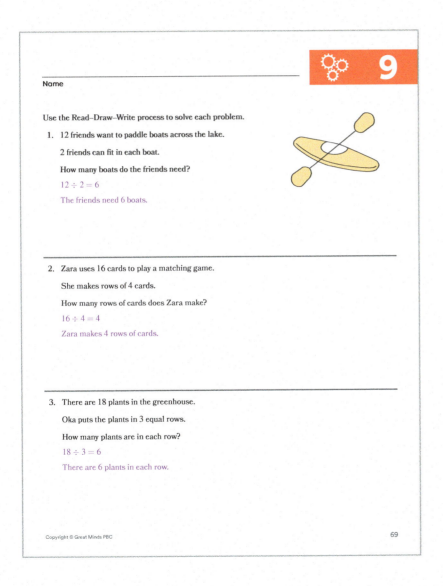

Name

Use the Read–Draw–Write process to solve each problem.

1. 12 friends want to paddle boats across the lake.

   2 friends can fit in each boat.

   How many boats do the friends need?

   $12 \div 2 = 6$

   The friends need 6 boats.

2. Zara uses 16 cards to play a matching game.

   She makes rows of 4 cards.

   How many rows of cards does Zara make?

   $16 \div 4 = 4$

   Zara makes 4 rows of cards.

3. There are 18 plants in the greenhouse.

   Oka puts the plants in 3 equal rows.

   How many plants are in each row?

   $18 \div 3 = 6$

   There are 6 plants in each row.

69

4. Ray puts 24 pictures in his scrapbook.

   He puts 4 pictures on each page.

   How many pages does Ray use?

   $24 \div 4 = 6$

   Ray uses 6 pages.

5. Mr. Davis sells peaches and apples at his fruit stand.

   a. Mr. Davis has 32 peaches to sell.

      He puts the peaches equally into 4 baskets.

      How many peaches are in each basket?

      $32 \div 4 = 8$

      There are 8 peaches in each basket.

   b. Mr. Davis has 27 apples to sell.

      He displays the apples in 3 equal rows.

      How many apples are in each row?

      $27 \div 3 = 9$

      There are 9 apples in each row.

Topic C
Properties of Multiplication

With the conceptual foundation for multiplication established in topic A and familiarity with arrays from topics A and B, the focus in topic C shifts to building fluency and developing multiplication strategies based on the commutative property of multiplication and the distributive property. Students explore the commutative property of multiplication for factors 2, 3, and 4 and use it as a multiplication strategy for working with larger factors (e.g., if I know $9 \times 4 = 36$, then I know $4 \times 9 = 36$).

Prior to topic C, rows in an array represent the unit and multiplication expressions are written using the convention of number of rows $\times$ the number in each row to help establish the meaning of multiplication and to maintain consistency. In topic C, as a bridge from the commutative property to the distributive property, students interchange rows and columns and refer to the number in each column as the number in each group. Students also shift to describing the size of each group, or its group size, rather than the number in each group.

To apply the distributive property, students partition arrays representing unknown multiplication facts into two smaller arrays representing known facts and then add the products of the smaller facts. This strategy is known by students as the break apart and distribute strategy. Students use number bonds and equations containing parentheses to show how the numbers were grouped and relate this work to counting the math way. As students progress through grade 3 and beyond, breaking apart a multiplication problem and distributing a factor to find partial products is a reliable strategy for multiplying large numbers and evaluating algebraic expressions.

In topics D and E, students apply the commutative and distributive properties as strategies to solve multiplication and division problems.

Progression of Lessons

Lesson 10

Demonstrate the commutative property of multiplication using a unit of 2 and the array model.

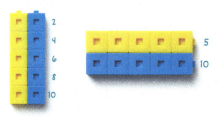

$$5 \times 2 = 2 \times 5$$

The commutative property of multiplication says that changing the order of the numbers in a multiplication fact does not change the product. I show this by rotating my array.

Lesson 11

Demonstrate the commutative property of multiplication using a unit of 4 and the array model.

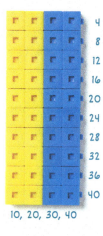

4
8
12
16
20
24
28
32
36
40

10, 20, 30, 40

I apply the commutative property of multiplication to change the order of the factors. This is a strategy I can use to find the product when I don't know the fact.

Lesson 12

Demonstrate the distributive property using a unit of 4.

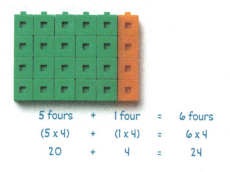

5 fours	+	1 four	=	6 fours
(5 × 4)	+	(1 × 4)	=	6 × 4
20	+	4	=	24

The column can represent the size of the group. When the unit is the same, composing smaller arrays into a larger array helps me to solve a larger multiplication fact.

Lesson 13

Demonstrate the commutative property of multiplication using a unit of 3 and the array model.

3
6
9
12
15
18
21
24
27
30

10, 20, 30

Arrays can be used flexibly to represent the commutative property of multiplication and to help solve problems.

Lesson 14

Demonstrate the distributive property using units of 2, 3, 4, 5, and 10.

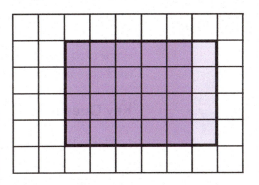

$$6 \text{ fours} = 5 \text{ fours} + 1 \text{ four}$$
$$6 \times 4 = (5 \times 4) + (1 \times 4)$$
$$24 = \quad 20 \quad + \quad 4$$

The break apart and distribute strategy helps me to break a large multiplication fact that I don't know into two smaller multiplication facts that I know. This helps me multiply.

Demonstrate the commutative property of multiplication using a unit of 2 and the array model.

Name _____

1. Draw an array to represent 2×3.

2. Explain how the array also shows 3×2.

The array also shows 3×2 because when I rotate it, there are 3 rows of 2. I did not add or take away any circles in the array. The product stayed the same.

91

Lesson at a Glance

Students develop an understanding of the unit of 2 and use the commutative property to show connections with other units. Students draw arrays to solve problems. This lesson formalizes the terms *rotate* and *commutative property of multiplication*.

Key Questions

- What changes when an array is rotated? What stays the same?
- How can the commutative property of multiplication help us learn new multiplication facts?

Achievement Descriptors

3.Mod1.AD1 Represent a multiplication situation with a model and **convert** between several representations of multiplication. (3.OA.A.1)

3.Mod1.AD5 Apply the commutative property of multiplication to multiply a factor of 2–5 or 10 by another factor. (3.OA.B.5)

Agenda

Fluency 10 min

Launch 10 min

Learn 30 min

- 5 twos or 2 fives?
- One Array, Two Equations
- Problem Set

Land 10 min

Materials

Teacher

- Envelopes (14)
- Equal Groups cards, Set A and Set B (in the student book)
- Interlocking cubes, 1 cm (10)

Students

- Envelopes of Equal Groups cards, Set A and Set B (1 per student pair)
- Sticky notes (6 per student pair)
- Interlocking cubes, 1 cm (10)

Lesson Preparation

- Tear out and cut apart the Equal Groups cards from the student book. One set per student pair and one set for the teacher are needed. Prepare seven envelopes with Set A cards and seven envelopes with Set B cards. Save cards for reuse in lesson 11.

- Prepare 10 interlocking cubes, 5 in one color and 5 in another color, per student and teacher.

Counting the Math Way by Tens and Threes

Students construct a number line with their fingers while counting aloud to build fluency with counting by tens, develop fluency with counting by threes, and develop a strategy for multiplying.

For each skip-count, show the math way on your own fingers while students count, but do not count aloud.

Let's count the math way by tens. Each finger represents 10.

Have students count the math way by tens from 0 to 100 and then back down to 0.

Show me 50.

(Students show 50 on their fingers using the math way.)

Have students count the math way by tens from 50 to 100 and then back down to 50.

Hands down. Now you count while I show my fingers. Ready?

Have students count forward and backward by tens, emphasizing counting on from 50.

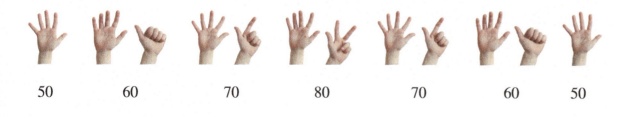

| 50 | 60 | 70 | 80 | 70 | 60 | 50 |

Now count the math way by threes. Each finger represents 3.

Have students count the math way by threes from 0 to 30 and then back down to 0.

Teacher Note

Have students watch and count aloud as you show the math way on your fingers. Continue to listen to student responses and be mindful of errors, hesitation, and lack of full-class participation. If needed, adjust the tempo or go back and forth to review numbers where you hear students falter. Expect a slower pace counting down.

Teacher Note

Threes facts are not formally taught until lesson 13. Therefore a productive struggle and slower pace is expected when counting by threes.

Sort: Relating Multiplication Models

Materials—S: Equal Groups cards, sticky notes

Students identify and sort models with units of 2, 3, 5, and 10 that represent the same multiplication expression and record the expression to build an understanding of multiplication.

Have students form pairs. Distribute either Set A or Set B cards and six sticky notes to each student pair. Have them sort the cards using the following procedure. Consider doing a practice round with students.

- Place all the cards faceup.
- Sort cards that model the same multiplication expression into a row.
- Use a sticky note to record the expression and place it next to the row of cards.

- Continue until all cards are sorted.

Circulate as students work and provide support as needed. If time permits, invite students to shuffle the cards and play again. Save the envelopes of cards. Students who had Set A today will get Set B in lesson 11. Students who had Set B today will get Set A in lesson 11.

Students engage in mathematical discourse to compare representations of equal groups.

Introduce the **Which One Doesn't Belong?** routine. Display the picture with four arrangements of objects and invite students to study them.

Give students 3 minutes to find a category in which three of the pictures belong, but a fourth picture does not.

When time is up, invite students to explain their chosen categories and to justify why one picture does not fit.

Highlight responses that emphasize reasoning about the arrangement of objects (e.g., equal groups, array, or scattered). Responses that reason about the representation of 5 and 2 (i.e., number of groups or number in each group) should also be highlighted.

Invite students to make connections and ask questions of their own, using precise language. Consider asking the following questions to guide the discussion:

- Why doesn't the picture of _____ belong?
- How are the sock and playing card pictures similar? How are they different?
- Does the picture of the eggs show 2 fives or 5 twos? Explain.
- What do all the pictures have in common?

Transition to the next segment by framing the work.

Today, we will describe an array in two ways.

Promoting the Standards for Mathematical Practice

When students explain their categories and analyze the categories of their peers, they are constructing viable arguments and critiquing the reasoning of others (MP3). Ask the following questions to promote MP3:

- Why are your categories correct? Convince the class.
- What parts of your classmates' categories do you question? Why?

Learn ⏱30

5 Twos or 2 Fives?

Materials—T/S: Cubes

Students create and rotate an array to represent related multiplication equations.

Partner students and prompt them to arrange the cubes to show 5 twos on their whiteboards.

What multiplication equation represents our cubes?

$5 \times 2 = 10$

Prompt students to make 2 fives. Each group of 5 cubes should be the same color.

What multiplication equation do our cubes represent now?

$2 \times 5 = 10$

If students have not already made an array, invite them to arrange their cubes to show 5 rows of 2, so that each column is the same color. Lead the class in a choral count by twos to find the total. Have students record the skip-count on their whiteboards as they say each number aloud. Then ask questions such as:

- What did we count by?
- How many twos did we count?
- What is the total of 5 twos?

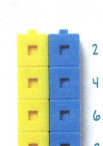

UDL: Representation

In topic C, multiplication is presented in a variety of ways. Students explore

- concrete interlocking cubes arranged in arrays,
- pictorial representations of arrays and tape diagrams, and
- equations using abstract numbers and symbols.

Students learn that they can rotate an array and represent a multiplication scenario with the same two factors in a different order, yet the product stays the same.

Prior knowledge is activated by using familiar models and by explicitly connecting to students' understandings about skip-counting. Finally, students practice applying multiplication in context, promoting transfer of learning.

Have students write a multiplication equation to represent 5 twos. After students have time to work, write $5 \times 2 = 10$.

Invite students to turn and talk about another way they could skip-count to find the total. Circulate and listen as students talk. Identify partners who discuss skip-counting by 5 to share their thinking with the class. Guide the discussion with questions such as:

What is another way we can count to get the total?

We can count by fives.

What part of the array is 5, the rows or the columns?

We turn, or rotate, our arrays to look at them another way. There are two ways to see the fives.

Model turning, or rotating, the array to show 2 rows of 5 and record the count at the end of each row. Prompt students to do the same.

Lead the class in a choral count by fives to find the total. Have students record the skip-count on their whiteboards as they say each number aloud. Then ask questions such as:

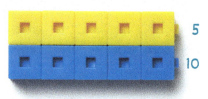

- What did we count by?
- How many fives did we count?
- What is the total of 2 fives?

Have students write a multiplication equation to represent 2 fives. After students have time to work, write $2 \times 5 = 10$.

Invite students to turn and talk to compare the arrays that represent 2 groups of 5 and 5 groups of 2. Guide students to notice that 5 rows of 2 is the same array as 2 rows of 5.

Why did the total number of cubes stay the same?

It is the same array. It was just rotated.

The position of the rows and columns changed, but the total number of cubes stayed the same.

Teacher Note

Rotating the array is a scaffold that is used to help students conceptualize commutativity. In lesson 11, students see that a single array can represent two equations by interpreting the row or the column as the number of groups.

How are the equations we wrote the same? How are they different?

They use the same numbers, but the position of the factors changes.

The product is the same.

Since $5 \times 2 = 10$ and $2 \times 5 = 10$, we can say $5 \times 2 = 2 \times 5$. We can change the order of the factors and have the same product. We call this the commutative property of multiplication.

One Array, Two Equations

Students draw an array and relate it to multiplication equations.

Direct students to complete the following prompts, one at a time, on their whiteboards. After each prompt, ask students to hold up their whiteboards. Scan student work and address any misconceptions.

Draw an array that represents 2×4.

Write two multiplication equations to describe the array.

Choose two samples of student work to share. Choose one that shows 2 rows of 4 and one that shows 4 rows of 2. Lead the class in a discussion to compare the two samples. Students should understand both arrays can represent 2×4. The rows and the columns can represent either the number of groups or the number in each group.

Repeat the process with 2×6 and 2×8 as time permits.

Direct students to the problem in their books. Encourage students to use the Read–Draw–Write process to solve the problem.

Teacher Note

Students should be familiar with the term *commutative property* and apply the concept of commutativity moving forward.

Language Support

Consider creating an anchor chart with students that highlights the commutative property of multiplication by showing that changing the order of the factors results in the same product.

5 rows of 2	2 rows of 5
$5 \times 2 = 10$	$2 \times 5 = 10$

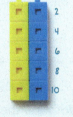 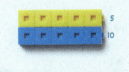

$5 \times 2 = 2 \times 5$

Commutative Property of Multiplication

Use the Read–Draw–Write process to solve the problem.

Pablo arranges his grapes into 7 rows. Each row has 2 grapes.
How many total grapes does Pablo have?

a. Draw an array to represent Pablo's grapes.

b. Write a multiplication equation to describe the array.

$7 \times 2 = 14$

c. Use the commutative property to write a different multiplication equation for the array.

$2 \times 7 = 14$

d. Complete the solution statement.

Pablo has __14__ total grapes.

Circulate as students work independently. When finished, ask students to compare their work with their partner. Guide a discussion using the following questions:

• How do your equations represent the word problem?

• How do your equations describe the array?

Problem Set

Differentiate the set by selecting problems for students to finish independently within the timeframe. Problems are organized from simple to complex.

Land 10

Debrief 5 min

Objective: Demonstrate the commutative property of multiplication using a unit of 2 and the array model.

Ask students to reflect on the work they did with cubes at the beginning of Learn.

What changed when we rotated the array? What stayed the same?

The rows became columns and the columns became rows. The total stayed the same.

Because of the commutative property of multiplication, what related fact do I know if I know 6×2?

2×6

And if I know what 9×2 is, I also know what?

2×9

How can the commutative property of multiplication help us learn new multiplication facts?

If we don't know a multiplication fact, we can switch the order of the factors to a fact we already know.

Exit Ticket 5 min

Provide up to 5 minutes for students to complete the Exit Ticket. It is possible to gather formative data even if some students do not complete every problem.

Sample Solutions

Expect to see varied solution paths. Accept accurate responses, reasonable explanations, and equivalent answers for all student work.

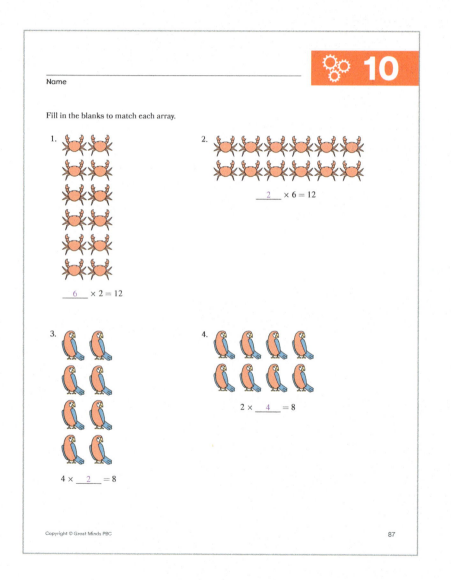

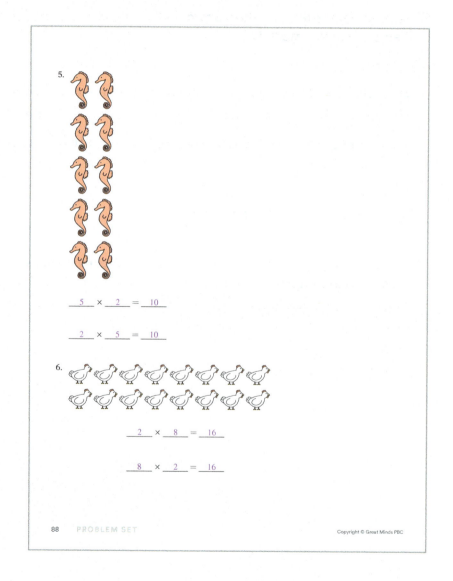

7. Jayla has 6 rows of 2 stickers.

a. Draw an array to represent the stickers.

b. Write two multiplication equations to describe the array.

$2 \times 6 = 12$

$6 \times 2 = 12$

8. Complete the equations.

$5 \times 2 = 2 \times \underline{5}$ $\underline{2} \times 8 = 8 \times 2$ $2 \times 10 = \underline{10} \times 2$ $2 \times \underline{9} = 9 \times 2$

9. a. Draw an array to show 2×4.

b. Explain how the array also shows 4×2.

The array also shows 4×2 because when I rotate it, there are 4 rows of 2. I did not add or take away any shapes in the array. The product stayed the same.

c. Complete the equation to show how 2 fours and 4 twos are related.

$\underline{2} \times \underline{4} = \underline{4} \times \underline{2}$

Demonstrate the commutative property of multiplication using a unit of 4 and the array model.

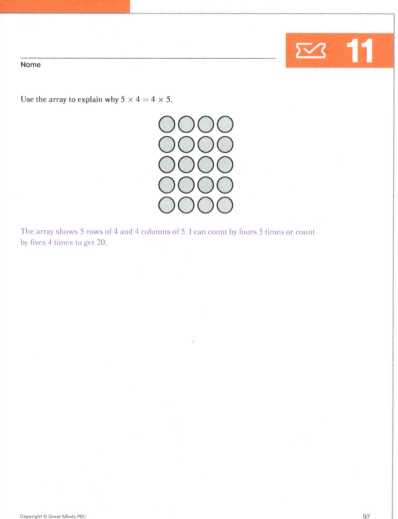

Lesson at a Glance

Students use their knowledge of units of 2 to learn units of 4. They skip-count rows and columns in an array to demonstrate the commutative property using equivalent expressions such as $4 \times 6 = 6 \times 4$. Students draw tape diagrams to represent the equal groups in an array.

Key Questions

- How does knowing $6 \times 4 = 24$ help us to know $4 \times 6 =$ _____ ?
- What changes and what stays the same when we use the commutative property?

Achievement Descriptors

3.Mod1.AD1 **Represent** a multiplication situation with a model and **convert** between several representations of multiplication. (3.OA.A.1)

3.Mod1.AD5 **Apply** the commutative property of multiplication to multiply a factor of 2–5 or 10 by another factor. (3.OA.B.5)

Agenda

Fluency 10 min

Launch 5 min

Learn 35 min

- Build an Array and Skip-Count by Fours
- Commutativity with Units of 4
- Tape Diagrams to Represent an Array
- Problem Set

Land 10 min

Materials

Teacher

- Interlocking cubes, 1 cm (40)

Students

- Envelopes of Equal Groups cards, Set A or Set B (1 per student pair)
- Sticky notes (6 per student pair)
- Interlocking cubes, 1 cm (40)

Lesson Preparation

- Gather envelopes of Equal Groups cards from lesson 10. If students had Set A cards in lesson 10, they should get Set B cards today (and vice versa).

- Prepare 40 interlocking cubes, 20 in one color and 20 in another color, per student and teacher.

Fluency 10

Counting the Math Way by Fives and Fours

Students construct a number line with their fingers while counting aloud to build fluency with counting by fives, develop fluency with counting by fours, and develop a strategy for multiplying.

For each skip-count, show the math way on your fingers while students count, but do not count aloud.

Let's count the math way by fives. Each finger represents 5.

Have students count by fives the math way from 0 to 50 and then back down to 0.

Show me 25.

(Students show 25 on their fingers by using the math way.)

Hands down. Now you count while I show my fingers. Ready?

Have students count forward and backward by fives, emphasizing counting on from 25.

| 25 | 30 | 35 | 40 | 35 | 30 | 25 |

Now count the math way by fours. Each finger represents 4.

Have students count the math way by fours from 0 to 20 and then back down to 0.

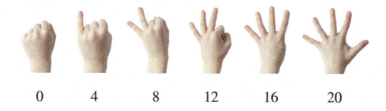

| 0 | 4 | 8 | 12 | 16 | 20 |

Differentiation: Support

If students need support with visualizing the value of each finger while counting by ones or by counting by a particular unit, consider using a number glove. Label the fingers by the unit by which you are counting.

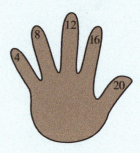

Teacher's right hand, open to face student.

Sort: Relating Multiplication Models

Materials—S: Equal Groups cards, sticky notes

Students identify and sort models with units of 2, 3, 5, and 10 that represent the same multiplication expression and record the expression to build an understanding of multiplication.

Have students form pairs. Distribute envelopes of Set B cards to student pairs who used Set A cards in lesson 10. Distribute envelopes of Set A cards to student pairs who used Set B cards in lesson 10. Provide six sticky notes to each student pair. Have students sort the cards using the following procedure.

- Place all the cards faceup.
- Sort cards that model the same multiplication expression into a row.
- Use a sticky note to record the expression and place it next to the row of cards.

- Continue until all cards are sorted.

Circulate as students work and provide support as needed. If time permits, invite students to shuffle the cards and play again.

Launch

 5

Students build on their knowledge of twos to see the relationship between twos and fours.

Show the picture of one car and ask students how many tires a car has. Model a think aloud to demonstrate.

> **A car has 4 tires. I don't see 4 tires in the picture, but I know there are 2 tires in the front and 2 tires in the back. I can count by twos to find the total number of tires. Since we know our twos well, let's count by twos to find the total number of tires. Ready?**

One at a time, show the pictures of 3, 6, and 9 cars. For each picture, ask students to tell how many tires there are and to explain how they counted efficiently. Reinforce strategies that include equal grouping and skip-counting.

Transition to the next segment by framing the work.

> **Today, we will use what we know about twos to multiply by fours.**

Teacher Note

Today's Launch prompts students to think about units of 2 within units of 4 to support the major work of Learn. The cars intentionally show 2 tires so that students do not count one-by-one and instead count by twos or fours to find the total number of tires.

Learn 35

Build an Array and Skip-Count by Fours

Materials—T/S: Cubes

Students relate units of 2 to units of 4 by using a cube array and skip-counting.

Use sticks of 2 to model joining the cubes to form a new unit of 4. Then build an array as shown. There should be spaces between the sticks. Prompt students to work with a partner to make the units of 4 with their cubes and to place them on their whiteboards.

Direct students to whisper-count by twos to help count by fours. Emphasize each multiple of 4 by saying it in a slightly louder voice.

Model pushing the rows together as you skip-count. Invite students to write the skip-count on their whiteboards as you model writing the skip-count next to each row.

2, 4, 6, 8, 10, 12, 14, 16, 18, 20, 22, 24, 26, 28, 30, 32, 34, 36, 38, 40

Invite students to turn and talk about how skip-counting by twos helps to skip-count by fours.

Work with your partner to find the value of each expression by using the array and skip-count.

Display the following expressions:

6×4 3×4 8×4 10×4

After providing time for students to work, discuss the strategies they used to multiply by 4.

How did you use the array and skip-count by fours to multiply?

To find 6×4, I used the array to count down 6 rows since I know each row has four. The skip-count shows the answer is 24.

I skip-counted by threes 4 times to find $3 \times 4 = 12$.

I started at the bottom because I know 10 fours is 40, and then I counted backward to find that 8 fours is 32.

Invite students to turn and talk about how the array and skip-count help to multiply by four.

Commutativity with Units of 4

Students skip-count by the rows and columns of an array to show the commutative property and use it to solve problems.

Look at the array of cubes. Where do you see 10 fours?

There are 10 rows of 4.

Where do you see 4 tens?

There are 4 columns of 10.

Let's show that. Write the skip-count at the bottom of each column as we count by tens.

10, 20, 30, 40

Did the total number of cubes change?

Since the total number of cubes did not change, we can say that 10 fours is the same number as 4 tens. What equation can we write to show that 10×4 is the same number as 4×10?

$10 \times 4 = 4 \times 10$

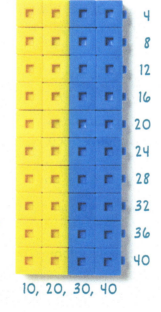

4
8
12
16
20
24
28
32
36
40

10, 20, 30, 40

Trace the 10 fours and 4 tens in the array while saying:

Yesterday we showed commutativity by rotating the array, but now we don't have to rotate it. We can see the 10 fours and 4 tens.

Continue to practice creating equations by using 5 fours and 4 fives, and 7 fours and 4 sevens.

Remove 2 rows from your array. How many fours do you have now?

8 fours

Promoting the Standards for Mathematical Practice

Students make use of structure (MP7) when they use concrete array models and abstract unit form expressions to represent and understand commutativity.

Ask the following questions to promote MP7:

- How can you use the different colors in each row to help skip-count by fours?

- How does what you know about 7×4 help you with knowing 4×7?

- How does what you know about 7 fours help you with knowing 4 sevens?

We can use commutativity to think about 8 fours as what?

4 eights

How could we find 4 eights if we don't know our eights?

We can count by fours 8 times.

4 eights is the same number as 8 fours, and I know 8 fours is 32.

Continue with a similar process to find 4×6 and 4×9. Support students in seeing that the rows and the columns can represent either the number of groups or the number in each group.

Invite students to turn and talk about how the commutative property can help solve unfamiliar facts.

Tape Diagrams to Represent an Array

Students draw tape diagrams to represent the rows and columns in an array.

Tell students that just like the cubes and arrays, tape diagrams can also show equal groups.

Let's relate our fours array to tape diagrams.

Show the 10 fours tape diagram.

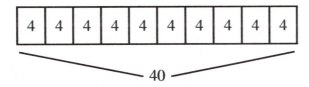

What multiplication equation can we use to represent this tape diagram, where the first factor is the number of groups?

185

UDL: Action & Expression

Continue to provide students with the option to use interlocking cubes as they transition to the tape diagram. While some students may prefer the use of tape diagrams immediately, others might favor the familiarity of the interlocking cubes until the factors get larger and they recognize the efficiency of tape diagrams. This allows students to express learning in flexible ways.

Show the 4 tens tape diagram.

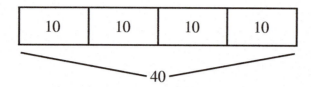

What multiplication equation can we use to represent this tape diagram, where the first factor is the number of groups?

How are the tape diagrams similar to the array we made to show $10 \times 4 = 4 \times 10$?

The array has 10 rows of 4 and the first tape diagram has 10 groups of 4. The array has 4 columns of 10 and the other tape diagram has 4 equal groups of 10.

One tape diagram shows 10 fours and the other tape diagram shows 4 tens. They both show 40 as the total, just like the array.

How are the tape diagrams different from the array?

We can show 10 fours and 4 tens in one array, but we can't show both in one tape diagram.

We see all 40 cubes in the array, but the tape diagrams only show the numbers that represent the number of rows or columns.

Ask students to draw tape diagrams to represent 5×4 and 4×5. Place emphasis on the tape diagrams being the same length because they have the same total.

Invite students to turn and talk about how tape diagrams relate to equal groups and arrays.

Problem Set

Differentiate the set by selecting problems for students to finish independently within the timeframe. Problems are organized from simple to complex.

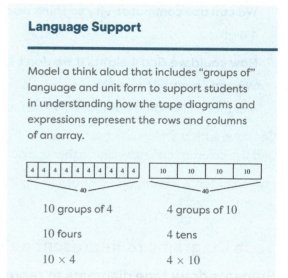

Language Support

Model a think aloud that includes "groups of" language and unit form to support students in understanding how the tape diagrams and expressions represent the rows and columns of an array.

10 groups of 4 | 4 groups of 10
10 fours | 4 tens
10×4 | 4×10

Land 10

Debrief 5 min

Objective: Demonstrate the commutative property of multiplication using a unit of 4 and the array model.

Direct students to look at problem 4 in their Problem Set. Have students think–pair–share to discuss the following:

How does knowing 6 fours is 24 help us know 4 sixes is 24?

If I know 6 fours, I know 4 sixes because it's the same array.

It's the same because I just switched the number of groups and the size of the group.

What model did we use today to show that $4 \times 6 = 6 \times 4$ **is true?**

We used an array and skip-counting.

Select a student who drew two arrays and a student who drew and labeled rows and columns in a 4-by-6 array to share their responses. Also select a student who drew and labeled rows and columns in a 6-by-4 array.

What changes and what stays the same when we use the commutative property?

The order of the factors changes.

The product stays the same.

Exit Ticket 5 min

Provide up to 5 minutes for students to complete the Exit Ticket. It is possible to gather formative data even if some students do not complete every problem.

Sample Solutions

Expect to see varied solution paths. Accept accurate responses, reasonable explanations, and equivalent answers for all student work.

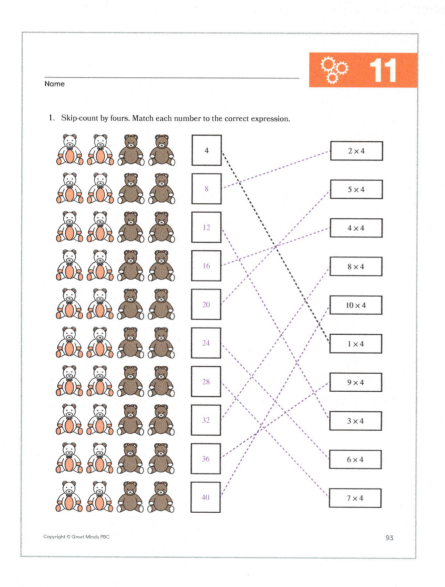

Name _____

11

1. Skip-count by fours. Match each number to the correct expression.

Complete the equations.

2. $6 \times 4 = \underline{} 24$

3. $\underline{} 6 \times 4 = 24$

4. $24 = \underline{} 4 \times 6$

5. $8 \times 4 = \underline{} 32$

6. $4 \times \underline{} 8 = 32$

7. $32 = \underline{} 4 \times 8$

8. Use the Read–Draw–Write process to solve the problem.

Miss Diaz has 5 tables.

There are 4 chairs at each table.

How many chairs are there altogether?

$5 \times 4 = 20$

There are ___20___ altogether.

9. Draw to show why the statement in the box is true. $4 \times 6 = 6 \times 4$

Sample:

◯◯◯◯◯◯ 6
◯◯◯◯◯◯ 12
◯◯◯◯◯◯ 18
◯◯◯◯◯◯ 24

4, 8, 12, 16, 20, 24

Label the tape diagrams to match their equations. Then complete the equations.

10.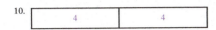

$2 \times 4 = \underline{8}$

11.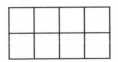

$4 \times 2 = \underline{8}$

12. How does the array show 2 fours and 4 twos?

The array has 2 rows of 4, which shows 2 fours. The array has 4 columns of 2, which shows 4 twos.

Demonstrate the distributive property using a unit of 4.

12

_____
Name

Shade the array to show two parts. Then fill in the blanks to describe the array.

Sample:

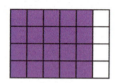

$\underline{\quad 5 \quad}$ fours $+$ $\underline{\quad 1 \quad}$ fours $= 6$ fours

$(\underline{\quad 5 \quad} \times 4) + (\underline{\quad 1 \quad} \times 4) = 6 \times 4$

$\underline{\quad 20 \quad} + \underline{\quad 4 \quad} = \underline{\quad 24 \quad}$

Lesson at a Glance

Students use the distributive property to compose known facts as a strategy to multiply 4 by factors such as 6, 7, 8, and 9. Terminology transitions from describing the number in a group to the size of the group. This lesson formalizes *parentheses* as a math term and symbol for showing groups in expressions and equations.

Key Questions

- How does knowing that large arrays are composed of smaller arrays help find the total?

- How do parentheses show the smaller arrays within a large array?

Achievement Descriptors

3.Mod1.AD6 Apply the distributive property to multiply a factor of 2–5 or 10 by another factor. (3.OA.B.5)

3.Mod1.AD8 Multiply and **divide** within 100 fluently with factors 2–5 and 10, recalling from memory all products of two one-digit numbers. (3.OA.C.7)

Agenda

Fluency 10 min

Launch 5 min

Learn 35 min

- Columns as the Size of the Group
- Compose a Large Array from Smaller Arrays
- Find Smaller Arrays inside a Large Array
- Problem Set

Land 10 min

Materials

Teacher

- Interlocking cubes, 1 cm (40)

Students

- Interlocking cubes, 1 cm (40)

Lesson Preparation

Prepare 40 interlocking cubes, 20 in one color and 20 in another color, per student and teacher.

Fluency

Whiteboard Exchange: Add and Subtract Within 100

Students identify and find the unknown in a number bond and equation to maintain work with addition and subtraction from grade 2.

Display the number bond.
Is the unknown a part or the total? Raise your hand when you know.

Wait until most students raise their hands, and then signal for students to respond.

> The total

> **How can you find the value for the unknown? Whisper your idea to your partner.**

Provide time for students to share with their partners.

> Add the parts.

Show the sample equation.

> **Find the unknown in the number bond and equation. Show your work.**

Give students time to work. When most students are ready, signal for students to show their whiteboards. Provide immediate and specific feedback. If students need to revise, briefly return to validate their corrections.

Show the unknown in the number bond and equation: 28.

Repeat the process with the following sequence:

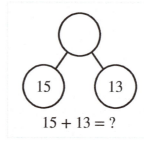

$15 + 13 = ?$

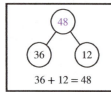

$36 + 12 = 48$

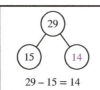

$29 - 15 = 14$

$57 - 22 = 35$

$23 + 68 = 91$

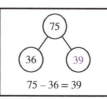

$75 - 36 = 39$

Teacher Note

Validate all ideas about equations that may not be displayed on the images. For example, the following number bond could represent $15 + ? = 29$ or $29 - ? = 15$.

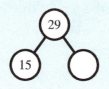

Teacher Note

Establish a signal (e.g., show me your whiteboards) to introduce a procedure for showing Whiteboard Exchange responses.

Practice with basic computations until students are accustomed to the procedure.

- What is $2 + 2$?

- What is $4 + 1$?

Prompt students to turn their whiteboards over to indicate to you that they are ready.

Counting the Math Way by Twos and Fours

Students construct a number line with their fingers while counting aloud to build fluency with counting by twos, develop fluency with counting by fours, and develop a strategy for multiplying.

For each skip-count, show the math way on your fingers while students count, but do not count aloud.

Let's count the math way by twos. Each finger represents 2.

Have students count by twos the math way from 0 to 20 and then back down to 0.

Show me 10.

(Students show 10 on their fingers the math way.)

Hands down. Now you count while I show my fingers. Ready?

Have students count forward and backward by twos, emphasizing counting on from 10.

| 10 | 12 | 14 | 16 | 14 | 12 | 10 |

Now count the math way by fours. Each finger represents 4.

Have students count the math way by fours from 0 to 40 and then back down to 0.

Choral Response: Commutative Property

Students find the product and use the commutative property to state a related equation, which develops use of the property as a strategy for multiplication.

Display $3 \times 10 =$ _____.

What is the product? Raise your hand when you know.

Wait until most students raise their hands, and then signal for students to respond.

30

Show the product: 30.

When I give the signal, switch the order of the factors and say the related equation.

$10 \times 3 = 30$

Show the related equation.

$$3 \times 10 = \underline{\quad 30 \quad}$$

$$10 \times 3 = 30$$

Repeat the process with the following sequence:

6×10	4×5	8×5	3×2	6×2

Launch 5

Students describe the composition of an array's parts and find the total.

Display the picture of the eggs. Provide students a few seconds to study the picture, then ask students to find the total number of eggs. Ask students to explain how they counted efficiently. Reinforce strategies that include equal grouping, such as skip-counting and seeing 2 fours + 2 fours.

Transition to the next segment by framing the work.

Today, we will use smaller multiplication facts to find the product of larger multiplication facts.

Columns as the Size of the Group

Materials—T/S: Cubes

Students use the columns as groups in an array to skip-count and write an equation.

Direct students to use one color of cubes to make an array with 5 columns of 4 on their whiteboards. Tell students that each column is a group. Then ask:

How many groups are there?

How many cubes are in each group?

Ask students to restate their answers using the following sentence frame: We have ___ groups of ____. Then invite the class to skip-count and use a finger to trace the columns of 4 to find the total. Record the count, as shown beneath the array.

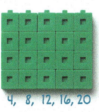

4, 8, 12, 16, 20

There are 4 cubes in each column. We can say 4 is the size of the group, because it is the number in each group.

Write an equation to describe the array. Use the number of groups as the first factor in the equation.

$5 \times 4 = 20$

Invite students to turn and talk to explain how the equation represents the array.

Compose a Large Array from Smaller Arrays

Students represent the distributive property with number bonds and equations with parentheses.

Use the following suggested sequence to model making a number bond with the cubes.

Show a stick of 1 four made from cubes that are a different color than the first array.

I have another array. This is 1 column of 4.

Display the two arrays and label each in unit form. Then complete the number bond as follows:

I want to make one large array using the two smaller arrays. Let's use a number bond to show how I can put the two arrays together.

Draw the number bond.

Combine the 5 fours array and the 1 four array into the total section of the number bond while asking:

5 fours and 1 four make how many fours?

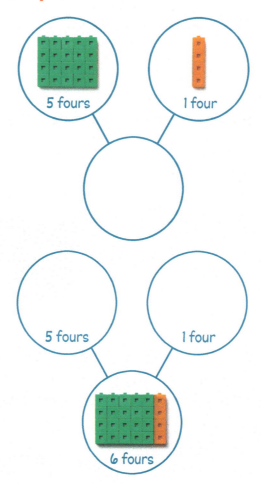

Teacher Note

In grade 2, students informally compose and decompose arrays and use number bonds to relate it to part–part–total thinking.

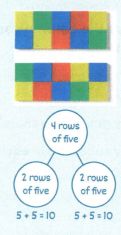

4 rows of five

2 rows of five   2 rows of five

5 + 5 = 10   5 + 5 = 10

Today's lesson is the first formal introduction to the distributive property in grade 3. To scaffold understanding, students are taught to compose smaller arrays to form a larger array. In lesson 14 students decompose larger arrays into smaller arrays and name it as the break apart and distribute strategy.

Label 6 fours in unit form.

We just showed that 5 fours and 1 four make 6 fours. Now our number bond is complete.

Direct students to make a number bond on their whiteboards and show 5 fours and 1 four as the parts. Have students move the parts together to show 6 fours as the total.

Connect the number bond work with students' experience counting the math way on their fingers.

Let's show what we did with the fours using our fingers. It is like counting the math way. Each finger is a unit of 4. Show me 6 fours on your fingers the math way. Now break the fours into two parts, 5 fours and 1 four, like the number bond.

Now put the fours back together. 5 fours and 1 four make 6 fours.

Invite students to turn and talk about how they made 6 fours with cubes and their fingers.

6 fours

5 fours

1 four

Direct students to move their 6 fours cube array out of the number bond to an empty space on their whiteboards. Support students in writing 5 fours $+$ 1 four $=$ 6 fours underneath their cube array. Guide students through the process of recording the remaining computations with the following possible prompts:

What multiplication expression represents 5 fours?

What multiplication expression represents 1 four?

5 fours	+	1 four	=	6 fours
(5×4)	+	(1×4)	=	6×4
20	+	4	=	24

Insert the $+$ and $=$ to form the equation as shown.

What multiplication expression represents 6 fours?

Insert parentheses around 5×4 and 1×4 while saying:

To help us solve, we can use parentheses to show our groups. Parentheses are symbols we use around an expression to show groups. They help us know what to do first when we solve. They show the order.

Find the value of the expressions in parentheses first. What is 5×4? 1×4?

What is $20 + 4$?

What multiplication problem did we find?

So 6 fours, or 6×4, is 24.

Leave the sequence displayed as a model for students in the next segment.

Using their cube arrays, invite students to break 6 fours into parts other than 5 and 1. Invite students to share other possible combinations, such as 3 fours and 3 fours, or 4 fours and 2 fours.

Teacher Note

Expression: An expression is a number or any valid combination of numbers and operations. An expression does not include an equal sign. For example:

$$5 \times 4$$

$$6 + 1$$

$$18 \div 3$$

Equation: An equation is a statement that two expressions are equal. For example:

$$5 \times 4 = 20$$

$$6 \times 4 = (5 \times 4) + (1 \times 4)$$

Language Support

Consider making the word *parentheses* visual by writing it as:

(PARENTHESES)

Highlight or write the ( ) in a different color.

Why do you think we started with 5 fours and 1 four? What is helpful about multiplying by 5 and 1 instead of 6?

If I don't know what 6 times 4 is, but I know how to multiply by 5, I can use my fives to help me. 6 fours is just 1 more four than 5 fours.

Note that naming this strategy comes later in lesson 14, after students have more experience using it.

Find Smaller Arrays Inside a Large Array

Students transition from concrete to pictorial representations with arrays by partitioning a large array into two smaller arrays.

Direct students to the problem in their books. Establish that the column is the group and ask students for the number of groups and the size of each group.

Complete the equations to describe the total number of squares in the array.

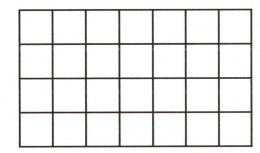

Sample: ___5___ fours + ___2___ fours = 7 fours

($\underline{}5\underline{} \times 4$) + ($\underline{}2\underline{} \times 4$) = 7×4

___20___ + ___8___ = ___28___

Guide students through the process of composing two smaller, more familiar multiplication facts to find 7 fours.

What are all the ways to make 7?

How can we compose 7 fours?

Make number bonds as students share. Invite students to think–pair–share about how they would compose 7 fours using familiar multiplication facts.

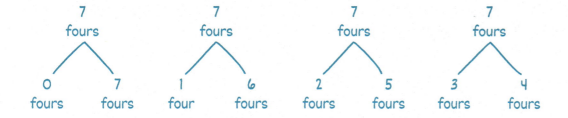

I can multiply by 5 and 2, so I would use 5 fours and 2 fours to make 7 fours.

Let's show this the math way. Each finger is a unit of four. Show me 7 fours on your fingers the math way. Now break the fours into 5 fours and 2 fours. 5 fours and 2 fours make 7 fours.

Allow students time to complete the problem. Consider encouraging students to compare their work with a partner.

Invite students to turn and talk about how the array and their written work are helpful when solving multiplication problems with unfamiliar factors.

Problem Set

Differentiate the set by selecting problems for students to finish independently within the timeframe. Problems are organized from simple to complex.

Land 10

Debrief 5 min

Objective: Demonstrate the distributive property using a unit of 4.

Facilitate a discussion about how breaking apart an array helps us multiply.

How does knowing that large arrays are composed of smaller arrays help find the total?

I can use facts that are easy for me to solve facts that I don't know.

How do parentheses show the smaller arrays within a large array?

The parentheses help me see the multiplication facts that represent the smaller arrays, or groups.

Exit Ticket 5 min

Provide up to 5 minutes for students to complete the Exit Ticket. It is possible to gather formative data even if some students do not complete every problem.

Sample Solutions

Expect to see varied solution paths. Accept accurate responses, reasonable explanations, and equivalent answers for all student work.

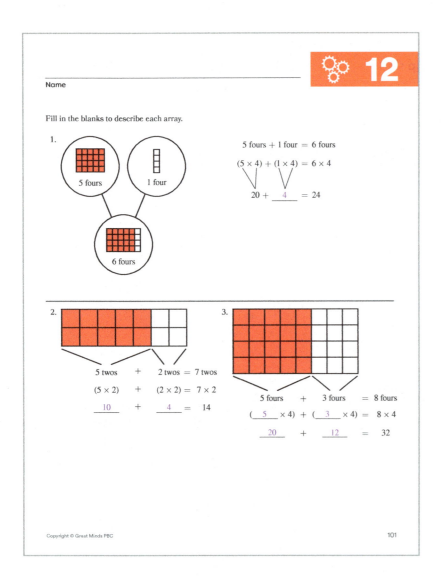

Name _____

12

Fill in the blanks to describe each array.

1.

5 fours   1 four   6 fours

5 fours + 1 four = 6 fours

$(5 \times 4) + (1 \times 4) = 6 \times 4$

$20 + \underline{4} = 24$

2.

5 twos   +   2 twos = 7 twos

(5×2) + $(2 \times 2) = 7 \times 2$

$\underline{10} + \underline{4} = 14$

3.

5 fours   +   3 fours   = 8 fours

$(\underline{5} \times 4) + (\underline{3} \times 4) = 8 \times 4$

$\underline{20} + \underline{12} = 32$

101

Shade each array to show two parts. Then fill in the blanks to describe each array.

4. Sample:

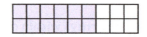

$\underline{6}$ twos + $\underline{3}$ twos = 9 twos

$(\underline{6} \times \underline{2}) + (\underline{3} \times \underline{2}) = \underline{18}$

$\underline{12} + \underline{6} = \underline{18}$

5. Sample:

$\underline{3}$ fours + $\underline{6}$ fours = 9 fours

$(\underline{3} \times \underline{4}) + (\underline{6} \times \underline{4}) = \underline{36}$

$\underline{12} + \underline{24} = \underline{36}$

102   PROBLEM SET

6. Oka puts her stickers in 5 columns of 4.

   a. Multiply to find Oka's total number of stickers.

   $$5 \times 4 = \underline{\ 20\ }$$

   Oka adds 2 more columns of 4.

   b. Add onto the array to show her new total.

   c. Fill in the blanks to represent 2 columns of 4.

   $$\underline{\ 2\ } \times 4 = \underline{\ 8\ }$$

   d. Find the total number of stickers Oka has. Add the products from parts (a) and (c).

   $$\underline{\ 20\ } + \underline{\ 8\ } = 28$$

   e. Write a multiplication equation that shows Oka's total.

   $$\underline{\ 7\ } \times \underline{\ 4\ } = \underline{\ 28\ }$$

13

Demonstrate the commutative property of multiplication using a unit of 3 and the array model.

Name

Draw to show that $3 \times 4 = 4 \times 3$. Explain how you know.

Sample:

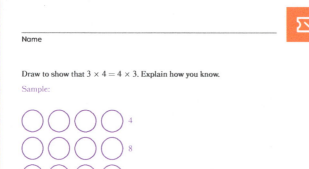

I see 3 rows of 4 and 4 columns of 3. My array didn't change, so $3 \times 4 = 4 \times 3$.

111

Lesson at a Glance

Students develop an understanding of the unit of 3 and use the commutative property to show connections with other units. They skip-count rows and columns in an array and write equations to find products of unfamiliar multiplication facts.

Key Questions

- What can we draw to show $3 \times 7 = 7 \times 3$?
- How does the commutative property of multiplication help us learn unfamiliar multiplication facts?

Achievement Descriptors

3.Mod1.AD1 Represent a multiplication situation with a model and **convert** between several representations of multiplication. (3.OA.A.1)

3.Mod1.AD5 Apply the commutative property of multiplication to multiply a factor of 2–5 or 10 by another factor. (3.OA.B.5)

Agenda

Fluency 10 min

Launch 5 min

Learn 35 min

- Build an Array and Skip-Count by Threes
- Commutativity with Units of 3
- Tape Diagrams to Represent an Array
- Problem Set

Land 10 min

Materials

Teacher

- Interlocking cubes, 1 cm (30)

Students

- Interlocking cubes, 1 cm (30)

Lesson Preparation

Prepare 30 interlocking cubes, of one color, per student and teacher.

Fluency 🔟

Whiteboard Exchange: Add and Subtract within 100

Students identify and find the unknown in an equation where the change is unknown to maintain work with addition and subtraction from grade 2.

Display $14 + ? = 29$.

Is the unknown a part or the total? Raise your hand when you know.

Wait until most students raise their hands, and then signal for students to respond.

A part

How can you find the value for the unknown? Whisper your idea to your partner.

Provide time for students to share with their partners.

Subtract the part from the total.

Add on to the part until you reach the total.

Find the unknown. Show your work.

Give students time to work. When most students are ready, signal for students to show their whiteboards. Provide immediate and specific feedback. If students need to revise, briefly return to validate their corrections.

$$14 + 15 = 29$$

Show the value of the unknown: 15.

Repeat the process with the following sequence:

$33 + 24 = 57$	$28 - 12 = 16$	$55 - 34 = 21$	$22 + 49 = 71$	$86 - 48 = 38$

Counting the Math Way by Threes and Fours

Students construct a number line with their fingers while counting aloud to develop fluency with counting by threes, fluency with counting by fours, and a strategy for multiplying.

For each skip-count, show the math way on your fingers while students count, but do not count aloud.

Let's count the math way by threes. Each finger represents 3.

Have students count the math way by threes from 0 to 30 and then back down to 0.

Now count the math way by threes from 0 to 15 and then back down to 0 with your partner. When you get to 15, give your partner a high five. When you get to 0, give your partner a fist bump.

0, 3, 6, 9, 12, 15, (high five), 15, 12, 9, 6, 3, 0 (fist bump)

Now let's count the math way by fours. Each finger represents 4.

Have students count the math way by fours from 0 to 40 and then back down to 0.

Now count the math way by fours from 0 to 20 and then back down to 0 with your partner. When you get to 20, give your partner a high five. When you get to 0, give your partner a fist bump.

0, 4, 8, 12, 16, 20, (high five), 20, 16, 12, 8, 4, 0 (fist bump)

Choral Response: Commutative Property

Students find the product and use the commutative property to state a related equation, which develops use of the property as a strategy for multiplication.

Display $9 \times 10 = ?$.

What is the product? Raise your hand when you know.

Wait until most students raise their hands, and then signal for students to respond.

90

> **Teacher Note**
>
> Introduce the partner component of this fluency to offer more movement and practice counting the math way. Encourage partners to count in unison.

Show the product: 90.

When I give the signal, switch the order of the factors and say the related equation.

$10 \times 9 = 90$

Show the related equation.

Repeat the process with the following sequence:

8×10	2×5	7×5	4×2	7×2

$9 \times 10 = \underline{\ 90\ }$

$10 \times 9 = 90$

Launch 5

Students count the total number of wheels on tricycles by skip-counting by threes.

Display the picture of the tricycle and ask how many wheels are on a tricycle.

One at a time, display the pictures of the tricycles. For each picture, ask students to tell how many total wheels there are and to explain how they counted efficiently. Reinforce strategies that include equal grouping and skip-counting.

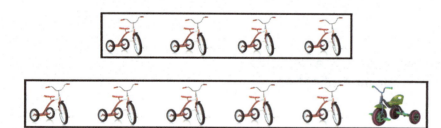

Transition to the next segment by framing the work.

Today, we will use strategies to help us multiply by 3.

Teacher Note

There is a specific significance to the last tricycle being different and facing the other way. This refers to the $5 + n$ strategy, or the break apart and distribute strategy. The intent is to informally connect the distributive property and counting the math way, which students do in the Learn section of this lesson.

Learn 35

Build an Array and Skip-Count by Threes

Materials—T/S: Cubes

Students connect units of 3 to multiplication equations by building a cube array and skip-counting.

Invite students to build an array on their whiteboards showing 10 units of 3, as you do the same.

When the array is finished, as a class, touch and whisper-count with an emphasis on the multiples of 3.

1, 2, **3**, 4, 5, **6**, 7, 8, **9**, 10, ..., 25, 26, **27**, 28, 29, **30**

Invite students to write the skip-count on their whiteboards as you model writing the skip-count next to each row.

Work with a partner to find the value of each expression using the array and skip-count.

Display the following expressions:

$$4 \times 3 \qquad 6 \times 3 \qquad 9 \times 3 \qquad 10 \times 3$$

After providing time for students to work, discuss the strategies they used to multiply by 3.

How did you use the array and skip-count by threes to multiply?

To find 4×3, I used the array to count down 4 rows since I know each row has 3.

I skip-counted by 3 six times to find $6 \times 3 = 18$.

I started at the bottom because I know 10 threes is 30. Then I counted backward to find 9 threes is 27.

Invite students to turn and talk about how arrays and skip-counting help to multiply by units of 3.

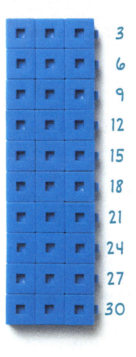

3
6
9
12
15
18
21
24
27
30

Language Support

Simultaneously whisper-counting while building the array is especially helpful for students who are still acquiring English. It helps them connect the concrete representation to the units we want them to see. If it is challenging for students to do these two tasks at the same time, consider building the array and then having students whisper-count as they touch each row.

Promoting the Standards for Mathematical Practice

As students recognize that skip-counting is a more efficient strategy than counting the individual objects in an array, they are looking for and expressing regularity in repeated reasoning (MP8).

Ask the following questions to promote MP8:

- When you look at each row of the array, is anything repeating? How can this repetition help you count more efficiently?

- How do you know that skip-counting by 3, instead of counting individual cubes, will help you find the total?

Commutativity with Units of 3

Students skip-count by the rows and columns of an array to show the commutative property and use it to solve problems.

Look at the array of cubes. Where do you see 10 threes?

There are 10 rows of 3.

Where do you see 3 tens?

There are 3 columns of 10.

Let's show that. Write the skip-count at the bottom of each column as we count by tens.

10, 20, 30

Did the total number of cubes change?

So 10 threes is the same number as 3 tens. What equation can we write to show that 10×3 is the same number as 3×10?

$10 \times 3 = 3 \times 10$

3
6
9
12
15
18
21
24
27
30

10, 20, 30

Continue to practice creating equations using 5 threes and 3 fives, and 7 threes and 3 sevens.

Remove 2 rows from your array. How many threes do you have now?

8 threes

We can use commutativity to think about 8 threes as what?

3 eights

How could we find 3 eights if we don't know our eights facts?

We can count by threes, 8 times.

3 eights is the same number as 8 threes, and I know 8 threes is 24.

I know 5 threes and 3 threes is 8 threes, so I can think $15 + 9 = 24$.

Continue with this sequence to find 3×6 and 3×9.

Invite students to turn and talk about how the commutative property can help solve unfamiliar facts.

UDL: Representation

Consider using color coding and annotation to emphasize that changing the order of the factors results in the same product. For example, write a multiplication equation using different colors for each factor and the product. Then rewrite the equation changing the order of factors. Label each factor and product.

$$3 \times 10 = 30$$
$$10 \times 3 = 30$$

Tape Diagrams to Represent an Array

Students draw tape diagrams to represent the rows and columns in an array.

Remind students that just like cubes and arrays, tape diagrams also show equal groups.

Let's relate our threes array to tape diagrams.

Direct students to draw a tape diagram to represent 10 threes.

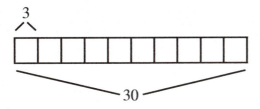

What multiplication equation can we use to represent this tape diagram, where the first factor is the number of groups?

Direct students to draw a tape diagram to represent 3 tens.

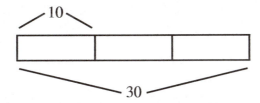

What multiplication equation can we use to represent this tape diagram, where the first factor is the number of groups?

Teacher Note

Labeling tape diagrams in this lesson is different from how they were labeled in lesson 11. In this lesson, the value of the unit is written on the outside of the tape diagram rather than inside each part. This is an example of flexibility in how tape diagrams are drawn. Students should label their tape diagrams in a way that makes sense to them.

How are the tape diagrams like the array we made to show $10 \times 3 = 3 \times 10$?

The array has 10 rows of 3, and the first tape diagram has 10 equal groups of 3. The array has 3 columns of 10, and the other tape diagram has 3 equal groups of 10.

One tape diagram shows 10 threes, and the other tape diagram shows 3 tens. They both show 30 as the total, just like the array.

We can change the order of the factors and get the same product because multiplication is commutative.

Ask students to draw an array and two tape diagrams to represent $3 \times 7 = 7 \times 3$. Place emphasis on the tape diagrams being the same length because they have the same total.

Invite students to turn and talk about how they could use the commutative property to find 3×6, 3×8, and 3×9.

Problem Set

Differentiate the set by selecting problems for students to finish independently within the timeframe. Problems are organized from simple to complex.

Land

Debrief 5 min

Objective: Demonstrate the commutative property of multiplication using a unit of 3 and the array model.

Ask students to look at problem 14 in their Problem Set. Have students think–pair–share to discuss the following:

What can we draw to show $3 \times 7 = 7 \times 3$?

We can draw arrays. The rows can be either the number of groups or the size of the groups.

We can use two tape diagrams to show that 3 sevens and 7 threes have the same product.

How does the commutative property help us learn unfamiliar multiplication facts?

We can switch the order of the factors.

If I know my twos, threes, fours, and fives, I also know any other multiplication fact that has a factor of 2, 3, 4, or 5.

Exit Ticket 5 min

Provide up to 5 minutes for students to complete the Exit Ticket. It is possible to gather formative data even if some students do not complete every problem.

Sample Solutions

Expect to see varied solution paths. Accept accurate responses, reasonable explanations, and equivalent answers for all student work.

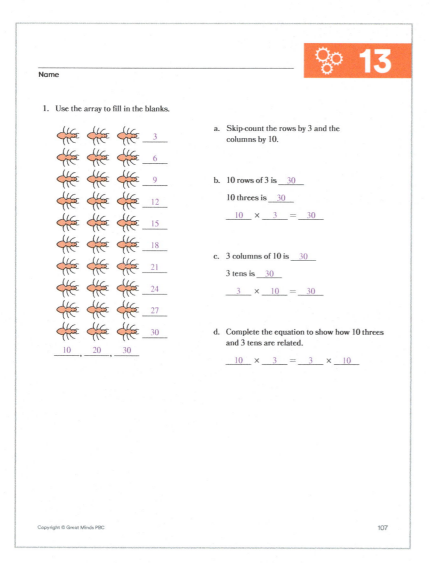

Name _____

13

1. Use the array to fill in the blanks.

   _3_

   _6_

   _9_

   _12_

   _15_

   _18_

   _21_

   _24_

   _27_

   _30_

_10_, _20_, _30_

a. Skip-count the rows by 3 and the columns by 10.

b. 10 rows of 3 is _30_

   10 threes is _30_

   _10_ × _3_ = _30_

c. 3 columns of 10 is _30_

   3 tens is _30_

   _3_ × _10_ = _30_

d. Complete the equation to show how 10 threes and 3 tens are related.

   _10_ × _3_ = _3_ × _10_

Complete the equations.

2. $4 \times 3 = $ _12_

3. _3_ $\times 4 = 12$

4. $12 = $ _4_ $\times 3$

5. $6 \times 3 = $ _18_

6. $3 \times$ _6_ $= 18$

7. $18 = $ _6_ $\times 3$

8. $7 \times$ _3_ $= 21$

9. $3 \times$ _7_ $= 21$

10. $21 = $ _3_ $\times 7$

11. $3 \times$ _9_ $= 27$

12. _9_ $\times 3 = 27$

13. $27 = 9 \times$ _3_

14. Draw to show why $3 \times 7 = 7 \times 3$.

Sample:

○○○○○○○ 7
○○○○○○○ 14
○○○○○○○ 21

3, 6, 9, 12, 15, 18, 21

15. Complete the table.

Label the tape diagram.	Complete the related equation.	Draw an array to match.
4 ⚙⚙⚙⚙ 12	__3__ × 4 = __12__	○○○○ ○○○○ ○○○○
3 ⚙⚙⚙ 12	4 × __3__ = __12__	○○○ ○○○ ○○○ ○○○

16. a. Circle two equations that show the commutative property of multiplication.

$\boxed{3 \times 5 = 5 \times 3}$

$4 \times 3 = 6 \times 2$

$3 \times 5 = 10 + 5$

$\boxed{8 \times 3 = 3 \times 8}$

b. Explain how the equations you circled in part (a) show the commutative property.

In the commutative property, the order of the factors can be switched.

$3 \times 5 = 15$ and $5 \times 3 = 15$, so $3 \times 5 = 5 \times 3$

$8 \times 3 = 24$ and $3 \times 8 = 24$, so $8 \times 3 = 3 \times 8$

LESSON 14

Demonstrate the distributive property using units of 2, 3, 4, 5, and 10.

✉ 14

Name _____

The array shows 7 broken apart to find 7 × 4. Fill in the blanks to match the array.

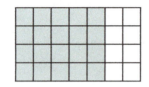

7 fours = ___5___ fours + ___2___ fours

7 × 4 = (___5___ × 4) + (___2___ × 4)

7 × 4 = ___20___ + ___8___

7 × 4 = ___28___

Lesson at a Glance

Students use the distributive property to multiply. Students break apart arrays, first concretely and then pictorially, into smaller arrays that guide them to use familiar facts. This lesson names the application of the distributive property as the break apart and distribute strategy.

Key Questions

- How does the break apart and distribute strategy help us multiply larger numbers more efficiently?
- When does it make sense to use the break apart and distribute strategy?

Achievement Descriptors

3.Mod1.AD6 Apply the distributive property to multiply a factor of 2–5 or 10 by another factor. (3.OA.B.5)

3.Mod1.AD.8 Multiply and **divide** within 100 fluently with factors 2–5 and 10, recalling from memory all products of two one-digit numbers. (3.OA.C.7)

Agenda

Fluency 10 min

Launch 5 min

Learn 35 min

- Compose a Large Array from Smaller Arrays
- Find Smaller Arrays Inside a Large Array
- Break Apart and Distribute
- Apply the Break Apart and Distribute Strategy to Context
- Problem Set

Land 10 min

Materials

Teacher

- Interlocking cubes, 1 cm (40)

Students

- Interlocking cubes, 1 cm (40)

Lesson Preparation

Prepare 40 interlocking cubes—20 in one color and 20 in another color—per student and teacher.

Fluency

Whiteboard Exchange: Add and Subtract Within 100

Students identify and find the unknown in an equation where the start is unknown to maintain work with addition and subtraction from grade 2.

Display $? + 15 = 29$.

Is the unknown a part or the total? Raise your hand when you know.

Wait until most students raise their hands, and then signal for students to respond.

A part

How can you find the value for the unknown? Whisper your idea to your partner.

Provide time for students to share with their partners.

Subtract the part from the total.

Add on to the part until you reach the total.

Find the unknown. Show your work.

Give students time to work. When most students are ready, signal for students to show their whiteboards. Provide immediate and specific feedback. If students need to revise, briefly return to validate their corrections.

$14 + 15 = 29$

Show the value of the unknown: 14.

Repeat the process with the following sequence:

$36 + 43 = 79$	$29 - 15 = 14$	$57 + 34 = 91$	$83 - 26 = 57$

Counting the Math Way by Threes and Fours

Students construct a number line with their fingers while counting aloud to develop fluency with counting by threes, fluency with counting by fours, and a strategy for multiplying.

For each skip-count, show the math way on your fingers while students count, but do not count aloud.

Let's count the math way by threes. Each finger represents 3.

Have students count the math way by threes from 0 to 30 and then back down to 0.

Now count the math way by threes from 0 to 15 and then back down to 0 with your partner. When you get to 15, give your partner a high five. When you get to 0, give your partner a fist bump.

0, 3, 6, 9, 12, 15, (high five), 15, 12, 9, 6, 3, 0 (fist bump)

Now let's count the math way by fours. Each finger represents 4.

Have students count the math way by fours from 0 to 40 and then back down to 0.

Now count the math way by fours from 0 to 20 and then back down to 0 with your partner. When you get to 20, give your partner a high five. When you get to 0, give your partner a fist bump.

0, 4, 8, 12, 16, 20, (high five), 20, 16, 12, 8, 4, 0 (fist bump)

Choral Response: Commutative Equations

Students find the product or factor and use the commutative property to state a related equation to develop use of the property as a strategy for multiplication.

Display _____ $\times 10 = 20$.

> **What is the value of the unknown factor? Raise your hand when you know.**

Wait until most students raise their hands, and then signal for students to respond.

> 2

Show the value of the unknown factor: 2.

> **When I give the signal, switch the order of the factors and say the related equation.**

> $10 \times 2 = 20$

Show the related equation.

$$\underline{\quad 2 \quad} \times 10 = 20$$

$$10 \times 2 = 20$$

Repeat the process with the following sequence:

___ $\times 10 = 50$	___ $\times 5 = 15$	___ $\times 5 = 30$	___ $\times 2 = 10$	___ $\times 2 = 16$

Launch 5

Students skip-count to find the total number of objects in an array.

One at a time, display the pictures of the apples. For each picture, invite students to turn and talk to find how many total apples there are and to explain how they counted efficiently. Reinforce strategies that include equal grouping and skip-counting.

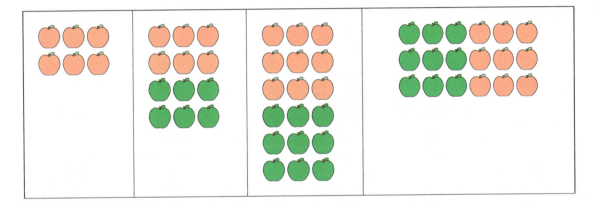

Transition to the next segment by framing the work.

Today, we will solve multiplication problems with larger factors by using multiplication facts we already know.

Learn 35

Compose a Large Array from Smaller Arrays

Materials—T/S: Cubes

Students concretely represent and solve an *array with unknown product* word problem by using the distributive property.

Direct students to problem 1. Instruct students to work with a partner to model the problem with cubes and answer each question.

1. Luke is selling vases of flowers.

   He arranges vases of pink flowers in 5 columns of three.

   He arranges vases of yellow flowers in 2 columns of three.

   a. How many total columns of vases are there?

      7 columns of vases

   b. How many vases of pink flowers are there?

      15 vases of pink flowers

   c. How many vases of yellow flowers are there?

      6 vases of yellow flowers

   d. How many total vases of flowers are there?

      21 total vases of flowers

Circulate and observe student work. Select a few students to share their work. Look for examples in which student models or drawings resemble the strategy as shown. Note that naming the strategy comes later in the lesson after students have more experience using it. If no one clearly uses this strategy, guide students to draw a number bond and write the equation with a sequence as in the following example.

5 threes + 2 threes = 7 threes

(5 x 3) + (2 x 3) = 7 x 3

15 + 6 = 21

> **Teacher Note**
>
> While teaching, consider including moments for students to pause and self-reflect. Pose a question to promote metacognition, such as:
>
> - What is helping you make sense of this strategy?
> - Where are you getting confused?
> - How does this strategy relate to other concepts you learned in math?
> - How can this strategy help you solve other problems?

Draw a number bond, and place a cube array for 5 threes and a cube array for 2 threes in the parts.

We can draw a number bond to show how we composed the columns of three. Show 5 columns of three to represent the vases of pink flowers and 2 columns of three to represent the vases of yellow flowers as the parts of the number bond.

Combine the 5 threes array and the 2 threes array into the total section of the number bond while saying the following:

5 threes and 2 threes are 7 threes. Show this with your cubes.

We can also show this with multiplication expressions.

Drag the 7 threes array out of the number bond to where there is room for writing the expressions. Instruct students to do the same.

What multiplication expression represents 5 threes?

Remember, to show how we group, we use parentheses. So we can put parentheses around 5 × 3.

Continue questioning and writing the expression for 2 threes and 7 threes as students do the same.

What is 5 × 3? 2 × 3?

What does the 15 represent in the problem?

What does the 6 represent in the problem?

What is 15 + 6?

So 7 × 3 is?

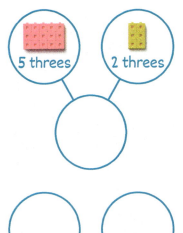

Teacher Note

The digital interactive Combine Arrays helps students visualize and work with properties of distributive multiplication.

Consider allowing students to experiment with the tool individually or demonstrating the activity for the whole class.

5 threes + 2 threes = 7 threes

(5 x 3) + (2 x 3) = 7 x 3

15 + 6 = 21

Invite students to complete the solution statement: Luke has a total of ____ vases of flowers.

> **The array and our written work help us see 7×3 as two simpler multiplication facts.**

Invite students to think-pair-share about how the array and the written work were helpful when completing a multiplication problem with unfamiliar factors.

> I know how to multiply by 5 and by 2, so those are easier facts for me. I do not know all my sevens yet.

> Instead of skip-counting by threes 7 times, I know $5 \times 3 = 15$, and then I can just count on 2 more threes.

Find Smaller Arrays Inside a Large Array

Students transition from concrete to pictorial representations of the break apart and distribute strategy with arrays by tracing and shading a large array on grid paper.

Direct students to problem 2. Invite students to shade a 7×3 array by using the following suggested prompts:

> **Each square on the grid paper is the same size as one of our cubes. We can use the grid paper to show our array. Put the 7×3 array you made with your cubes on top of the grid to line up with the squares.**

> **Carefully pick up one of the pink threes. Heavily shade in the squares that are under the three cubes so they are dark.**

2.

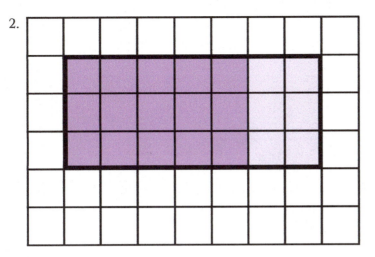

Guide students to repeat shading columns of 3 until 5 columns are heavily shaded to represent the vases of pink flowers.

Repeat for the 2 threes representing the vases of yellow flowers. Shade these squares lightly for contrast.

Trace the lines around the entire array to create a border.

Invite students to think–pair–share about how the array represents the problem.

The 5 columns of three represent the 15 vases of pink flowers. The 2 columns of three represent the 6 vases of yellow flowers. The whole array represents the 21 total vases of flowers.

Break Apart and Distribute

Students decompose an array pictorially into smaller arrays and name the strategy as the break apart and distribute strategy.

Let's draw an array and break it apart to help us multiply.

Direct students to problem 3. Invite students to trace an array of 6 columns of 4 on the grid. Then instruct them to shade 5 fours heavily and 1 four lightly.

3.

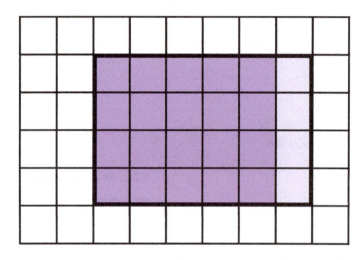

$$\underline{6} \text{ fours } = \underline{5} \text{ fours } + \underline{1} \text{ four}$$

$$\underline{6} \times 4 = (\underline{5} \times 4) + (\underline{1} \times 4)$$

$$\underline{24} = \underline{20} + \underline{4}$$

Guide students through completing the computation by using a sequence such as the following:

How many fours are in the whole array?

How many fours are heavily shaded?

How many fours are lightly shaded?

What multiplication expression represents 6 fours?

Continue guiding students through the strategy by asking for the multiplication expression that represents 5 fours and 1 four. Use parentheses to show the grouping.

Find the value of the expressions in parentheses first before adding. What is 5×4? 1×4?

What is $20 + 4$?

So 6×4 is?

This strategy is called the break apart and distribute strategy. We use it to break apart larger factors, like the 6, into more familiar factors to help us multiply. There is usually more than one way we can break apart the factors. We want to choose a way that is efficient because it uses facts that we know well.

Use a similar sequence to find 8×3 and 9×4 as time allows. Be less explicit with what smaller factors students use to break apart the larger factor.

Invite students to turn and talk about which problems they would use the break apart and distribute strategy to solve.

Language Support

To help students solidify their understanding of the break apart and distribute strategy, revisit the number bond. Build an array. Show the decomposition and composition so each part can be analyzed.

Differentiation: Challenge

Consider challenging students to identify as many ways as possible to break apart an array that represents a fact. Ask them to explain how some combinations are more efficient than others and how they decide how to break apart the array.

Apply the Break Apart and Distribute Strategy to Context

Students apply the break apart and distribute strategy to find a product.

Direct students to problem 4.

4. A shoe store has a display of 9 stacks of 3 boxes of shoes.

   Some stacks of 3 boxes are on sale.

   The rest of the stacks are not on sale.

   a. Draw one possible combination of stacks that are on sale and stacks that are not.

   b. Write expressions and equations to represent your drawing and how many boxes there are altogether.

   c. Write a solution statement to describe how many boxes there are altogether.

   Sample:

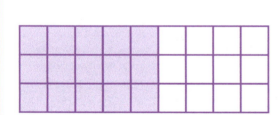

$$9 \text{ threes} = 5 \text{ threes} + 4 \text{ threes}$$
$$9 \times 3 = (5 \times 3) + (4 \times 3)$$
$$27 = 15 + 12$$

There are 27 boxes altogether.

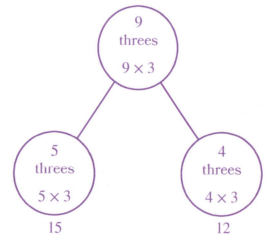

$$15 + 12 = 27$$

There are 27 boxes altogether.

Promoting the Standards for Mathematical Practice

Students reason quantitatively and abstractly (MP2) as they decontextualize the word problem by drawing, writing, and manipulating mathematical representations of the story.

Ask the following questions to promote MP2:

- How does the shaded part of your array relate to the original problem?

- How do your expressions and equations represent the stacks that aren't on sale and the stacks that are on sale?

Provide students time to work. Circulate and observe student work. Look for work that models the break apart and distribute strategy.

Invite a few students to share their work. As each student shares, ask questions to elicit their thinking and clarify the strategy. Ask the class questions that invite students to make connections between different representations that show the break apart and distribute strategy.

Invite students to turn and talk about how the drawings and written work used to show the break apart and distribute strategy are helpful when solving multiplication problems with unfamiliar factors.

Problem Set

Differentiate the set by selecting problems for students to finish independently within the timeframe. Problems are organized from simple to complex.

Land 10

Debrief 5 min

Objective: Demonstrate the distributive property using units of 2, 3, 4, 5, and 10.

Facilitate a discussion of the benefits and limitations of the break apart and distribute strategy to prepare students for choosing an appropriate strategy on future problems.

How does the break apart and distribute strategy help us multiply larger numbers more efficiently?

I can break apart the problem into simpler multiplication facts that I know so I can multiply quickly.

When does it make sense to use the break apart and distribute strategy?

It makes sense to use it when the problem is a fact I don't know.

I can use it when the factors are hard to skip-count.

It is helpful to use when one factor is large.

Exit Ticket 5 min

Provide up to 5 minutes for students to complete the Exit Ticket. It is possible to gather formative data even if some students do not complete every problem.

Sample Solutions

Expect to see varied solution paths. Accept accurate responses, reasonable explanations, and equivalent answers for all student work.

14

Name _____

Fill in the blanks to describe each array.

1.

9 threes

5 threes   4 threes

9 threes = 5 threes + 4 threes

$9 \times 3 = (5 \times 3) + (4 \times 3)$

$27 = \underline{\ 15\ } + \underline{\ 12\ }$

2.

8 threes = 5 threes + 3 threes

$8 \times 3 = (5 \times 3) + (\underline{\ 3\ } \times 3)$

$24 = \underline{\ 15\ } + \underline{\ 9\ }$

3.

7 threes = 5 threes + 2 threes

$7 \times 3 = (\underline{\ 5\ } \times 3) + (\underline{\ 2\ } \times 3)$

$21 = \underline{\ 15\ } + \underline{\ 6\ }$

4. Show two different ways to make 6 threes. Shade the arrays and complete the equations.

Sample:

6 threes = \underline{\ 3\ } threes + \underline{\ 3\ } threes

$6 \times 3 = (\underline{\ 3\ } \times 3) + (\underline{\ 3\ } \times 3)$

$18 = \underline{\ 9\ } + \underline{\ 9\ }$

6 threes = \underline{\ 2\ } threes + \underline{\ 4\ } threes

$6 \times 3 = (\underline{\ 2\ } \times 3) + (\underline{\ 4\ } \times 3)$

$18 = \underline{\ 6\ } + \underline{\ 12\ }$

117

118   PROBLEM SET

5. Liz buys a box of yogurt cups.

The box has 7 columns. There are 4 yogurt cups in each column.

There are 2 columns of lemon yogurt, and the rest are peach.

a. Draw the array of lemon and peach yogurt cups.

b. Fill in the blanks to find the total number of yogurt cups.

$$(\underline{2} \times 4) + (5 \times 4) = \underline{8} + \underline{20}$$

$$7 \times 4 = \underline{8} + \underline{20}$$

$$7 \times 4 = \underline{28}$$

6. There are 9 columns of 4 mailboxes in the office.

Some of the columns have gray mailboxes.

The rest of the columns have white mailboxes.

a. Draw to show one possible combination of gray and white mailboxes.

Sample:

b. Write equations to represent your drawing.

$$(9 \times 4) = (4 \times 4) + (5 \times 4)$$
$$36 = 16 + 20$$

c. How many mailboxes are there altogether?

There are 36 mailboxes altogether.

Topic D
Two Interpretations of Division

Topic D lessons provide students with multiple opportunities to build understanding of the two interpretations of division: partitive (number of groups known) and measurement (size of each group known). Students expand their foundational understanding of division from topic B by establishing the unknown factor in a context and constructing a tape diagram to match. Relating the quotient to an unknown factor problem supports students as they progress toward fluency with multiplication and division facts within 100. As students contextually explore measurement and partitive division, they analyze the differences and relate the solution back to the original problem.

The relationship between the number of groups, the size of each group, and the total continues to be developed as students apply their understanding to tape diagrams. Experience with each interpretation of division helps students analyze and plan for the construction of a division tape diagram before they start drawing. Tape diagrams for measurement division tend to be more challenging for students because they must anticipate how many equal partitions to make as they construct the tape diagram. When students understand the differences between the two interpretations of division, they can accurately draw models that represent problems and use them to solve the one-step problems included in this topic and the more complex problems they encounter later.

In topic E, students apply their understanding of the two interpretations of division to the distributive property and to multi-step problems. In module 3, students continue to experience examples of the two interpretations of division with larger factors.

Progression of Lessons

Lesson 15

Model division as an unknown factor problem.

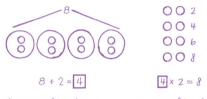

$8 \div 2 = \boxed{4}$

4 cups of iced tea have lemon slices.

$\boxed{4} \times 2 = 8$

4 cups of iced tea have lemon slices.

To find the quotient in a division problem, it can be helpful to think about multiplication. Finding the quotient is like finding an unknown factor in a multiplication problem.

Lesson 16

Model the quotient as the number of groups using units of 2, 3, 4, 5, and 10.

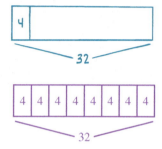

Sometimes when I divide, I know the total and the size of each group. The unknown is the number of groups. I can draw a tape diagram to help me see a solution strategy. I can finish the tape diagram after I find the quotient and know how many groups there are.

Lesson 17

Model the quotient as the size of each group using units of 2, 3, 4, 5, and 10.

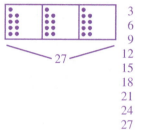

Sometimes when I divide, I know the total and the number of groups. Drawing a tape diagram when I know the number of groups is different than when I know the size of each group. I draw equal parts to represent the number of groups and then equally share among the groups. A skip-count can help track how many of the total I have shared.

Lesson 18

Represent and solve measurement and partitive division word problems.

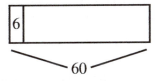

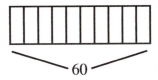

When I solve a division word problem, I need to make sure my drawing and equation correctly show the unknown, whether it's the number of groups or the size of each group. The Read–Draw–Write process helps me to make sense of a problem and to see the solution path.

15

Model division as an unknown factor problem.

✉ **15**

Name _____

Carla has 12 grapes. She gives 4 grapes to each friend. How many friends get grapes?

a. Draw a picture to represent the problem.

Sample:

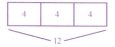

b. Complete the equations to find the unknown.

$3 \times 4 = 12$

$12 \div 4 = 3$

c. _3_ friends get grapes.

Lesson at a Glance

Students relate unknown factor equations to division. They represent division problems using equal groups, arrays, and tape diagrams. This lesson formalizes the term *quotient*.

Key Questions

- What is the relationship between the quotient in division and the unknown factor in a related multiplication equation?
- What helps you identify the unknown as either the number of groups or the size of each group?

Achievement Descriptors

3.Mod1.AD2 **Represent** a division situation with a model and **convert** between several representations of division. (3.OA.A.2)

3.Mod1.AD4 **Determine** the unknown number in a multiplication or division equation involving factors and divisors 2–5 and 10. (3.OA.A.4)

3.Mod1.AD7 **Represent** and **explain** division as an unknown factor problem. (3.OA.B.6)

Agenda

Fluency 10 min

Launch 10 min

Learn 30 min

- Express Division as an Unknown Factor Problem
- Represent Measurement Division with a Tape Diagram
- Problem Set

Land 10 min

Materials

Teacher

- None

Students

- None

Lesson Preparation

None

Students tell time on a digital clock to the nearest five minutes, using picture clues to distinguish between a.m. and p.m., to maintain work with time from grade 2.

After asking each question, wait until most students raise their hands, and then signal for students to respond.

> **Raise your hand when you know the answer to each question. Wait for my signal to say the answer.**

Display the picture that shows 2:00.

> **What time does the clock show?**
>
> 2:00
>
> **Is it 2:00 a.m. or p.m.?**
>
> p.m.

Repeat the process with the following sequence:

1:00 a.m.   7:30 a.m.   12:30 p.m.   8:15 a.m.   6:35 p.m.

Counting the Math Way by Twos and Threes

Students construct a number line while counting aloud to build fluency with counting by twos and threes and develop a strategy for multiplying.

For each skip-count, show the math way on your fingers while students count, but do not count aloud.

Let's count the math way by twos. Each finger represents 2.

Have students count the math way by twos from 0 to 20 and then back down to 0.

Hands down. Now you count aloud while I show the count on my fingers. Ready?

Lead students to count aloud forward and backward by twos, emphasize counting on from 10.

Now let's count the math way by threes. Each finger represents 3.

Have students count the math way by threes from 0 to 30 and then back down to 0.

Show me 15.

(Students show 15 on their fingers the math way.)

Have students count the math way by threes from 15 to 30 and then back down to 15.

Whiteboard Exchange: Divide Equal Groups

Students write division equations to describe an equal-groups picture to build fluency with two interpretations of division and associated terminology.

After asking each question, wait until most students raise their hands, and then signal for students to respond.

Raise your hand when you know the answer to each question. Wait for my signal to say the answer.

Display the picture of the apples.

What is the total number of apples?

10

What is the total number of groups?

2

What is the size of each group?

5

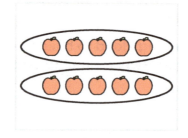

After each prompt for a written response, give students time to work. When most students are ready, signal for students to show their whiteboards. Provide immediate and specific feedback. If students need to revise, briefly return to validate their corrections.

Write a division equation where the answer is the size of each group.

Show the equation: $10 \div 2 = 5$.

Write a division equation where the answer is the number of groups.

Show the equation: $10 \div 5 = 2$.

Repeat the process with the following sequence:

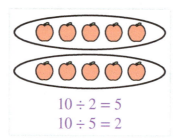

$10 \div 2 = 5$
$10 \div 5 = 2$

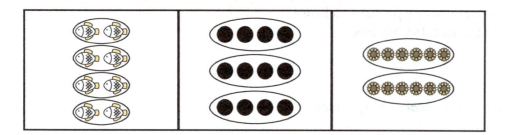

Launch ⑩

Students relate division to multiplication by seeing division as an unknown factor problem.

Direct students to problem 1 in their books. Ask students to write two multiplication equations to represent each picture. Then ask students to draw a box around the factor that represents the number of groups in each equation. Clarify that for picture c, the rows are the number of groups. For picture d, the columns are the number of groups.

1.

Picture	Multiplication Equations	Division Equation
a.	$\boxed{2} \times 3 = 6$ $3 \times \boxed{2} = 6$	$6 \div 3 = \boxed{2}$
b.	$\boxed{3} \times 4 = 12$ $4 \times \boxed{3} = 12$	$12 \div 4 = \boxed{3}$
c.	$\boxed{4} \times 2 = 8$ $2 \times \boxed{4} = 8$	$8 \div 2 = \boxed{4}$

d.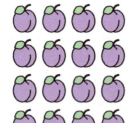

$$\boxed{4} \times 5 = 20$$
$$5 \times \boxed{4} = 20$$

$$20 \div 5 = \boxed{4}$$

For each picture, ask the corresponding question below. Then ask students to write a division equation for each picture and to draw a box around the answer.

a. How many threes are in 6?

b. How many fours are in 12?

c. How many twos are in 8?

d. How many fives are in 20?

Are the answers to each division equation the number of groups or the size of each group?

The number of groups

Facilitate a discussion about the relationship between the multiplication and division equations using the following questions:

Let's compare the three equations for the cherry picture. What is the same and different about them?

All the equations use 2, 3, and 6, but they are in different places.

The 2 represents the number of groups and is boxed in each of the equations.

What relationships do you notice between the equations you wrote for the oranges, pears, and plums?

The equations for the other fruits also use the same numbers but in different places.

The boxed numbers are the number of groups.

The three equations in each set are related because they describe the same picture and use the same three numbers.

Invite students to turn and talk about how knowing multiplication facts can help solve division problems.

Transition to the next segment by framing the work.

Today, we will use the relationship between multiplication and division to solve problems.

Express Division as an Unknown Factor Problem

Students represent an *equal-groups with number of groups unknown* word problem and solve it with a related multiplication equation.

Direct students to problem 2. Prompt students to use the Read–Draw–Write process and select their own strategy to solve.

Use the Read–Draw–Write process to solve the problem.

2. Mia puts lemon slices into cups of iced tea.

   She has 8 lemon slices.

   She puts 2 lemon slices in each cup.

   How many cups of iced tea have lemon slices?

   4 cups of iced tea have lemon slices.

Circulate and observe student work. Select a few students to share their work. Look for a student's drawing that highlights equal groups and another that highlights arrays.

Use the following questions to lead a discussion:

What is unknown: the number of groups or the size of the group? How do you know?

The number of groups is unknown. We know the total number of lemon slices and how many are in each cup, but we don't know how many cups there are.

Display one student work sample of an equal-groups model and one of an array model.

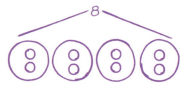

$$8 \div 2 = \boxed{4}$$

4 cups of iced tea have lemon slices.

$$\boxed{4} \times 2 = 8$$

4 cups of iced tea have lemon slices.

Where do you see the unknown in the equal-groups model? Where do you see the unknown in the array?

In the equal-groups model, the large circles represent the cups.

In the array, each row represents a cup.

Write _____ $\times 2 = 8$.

We need to find the unknown factor. How does this equation represent both the array and equal groups?

We know that 2 is the size of the group and 8 is the total. We are trying to find how many groups there are.

Promoting the Standards for Mathematical Practice

Students model with mathematics (MP4) when they solve division problems using familiar models of equal groups, arrays, tape diagrams, and equations.

Ask the following questions to promote MP4:

- What key ideas from problem 2 do you need to make sure are included in your drawing?

- What does each number in your equation tell you about the iced tea and the lemons?

Point to the blank in the equation and ask students to name the unknown factor. Write 4 in the equation and box it. Invite students to write the completed multiplication equation. Then ask guiding questions such as:

What does 4 represent in the problem?

The number of cups

Do the cups represent the number of groups or the size of the group?

The number of groups

What division equation can be used to find the number of cups?

$8 \div 2 =$ _____

Write the equation $8 \div 2 =$ _____ and ask students for the answer. Write 4 as the quotient in the equation. Ask students to write the division equation and explain how the division equation represents the problem.

When we divide, we call the answer the quotient. Let's box the quotient. Which number is the quotient?

4

Direct students to draw a box around 4 in the equation and to write the word *quotient*.

How is the division equation similar to the equation with the unknown factor?

The unknown in both equations is the number of groups.

Invite students to turn and talk about how the multiplication and division equations are related.

Copyright © Great Minds PBC

Teacher Note

There are two distinct interpretations of division: *partitive* and *measurement*.

In partitive division, the total and the number of groups are known and the size of each group is unknown. Partitive division asks the question, What is the size of each group?

In measurement division, the total and size of each group are known and the number of groups is unknown. Measurement division asks the question, How many groups are there?

Students are not expected to know the terms *partitive division* and *measurement division*, but they are expected to identify which number in a situation or equation represents the number of groups and which represents the size of each group.

Represent Measurement Division with a Tape Diagram

Students reason about, represent, and solve *equal-groups with number of groups unknown* **word problems using array and tape diagram models.**

Model creating the tape diagram shown.

Watch as I show how we can represent this problem another way using a tape diagram.

We know there are 8 total lemon slices. I will draw a tape diagram and label the whole tape with an 8 to represent the 8 lemon slices.

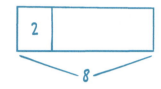

There are 2 lemon slices in each cup of iced tea. The total 8 will be made up of equal parts of 2. I will draw one part and label it 2, to represent the 2 lemon slices. The tape diagram shows me that I know the group size. I need to find the number of groups.

We can find the number of groups by asking ourselves, how many twos are in 8?

Write _____ $\times 2 = 8$.

What strategy can we use to find how many twos are in 8?

Skip-count by twos until we get to 8.

Let's skip-count by twos until we say 8.

Chorally count by twos and track the count the math way with fingers.

How many twos are in 8?

4

Let's show this on the tape diagram.

Complete the tape diagram and ask students to skip-count by twos as you touch each part in the tape diagram.

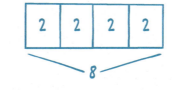

What do the 4 groups represent in the problem?

The 4 groups represent the 4 cups of iced tea.

Invite students to think–pair–share to relate the tape diagram to the array and equal-groups models with questions such as the following:

How is the tape diagram similar to the array and equal-groups models? How is it different?

The tape diagram, equal groups, and the array models all represent the problem.

The tape diagram has numbers to tell the size of the group. The array and equal groups have dots and circles.

How does the tape diagram represent $4 \times 2 = 8$?

There are 4 groups with 2 in each group. The total is 8.

How does the tape diagram represent $8 \div 2 = 4$?

The total of 8 is divided into equal groups of 2. There are 4 groups of 2.

Write a solution statement to answer the question.

Invite one student to share their solution statement.

Teacher Note

Throughout module 1, students are encouraged to complete tape diagrams by labeling all components: the total, the number of groups, and the number in each group. This supports their understanding of the problems and validates their solutions.

In module 3, as they gain deeper conceptual understanding and the size of divisors increases, students transition to using a symbol to identify the unknown.

Direct students to problem 3. Prompt students to use the Read–Draw–Write process to solve.

Use the Read–Draw–Write process to solve the problem.

3. A fruit stand has a total of 40 plums on a table. They are arranged in rows.

There are 5 plums in each row.

How many rows of plums are there?

a. Draw an array to represent the problem.

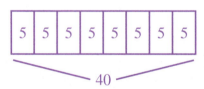

b. Draw a tape diagram to represent the problem.

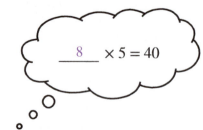

Complete the equations to find the unknown.

$$\underline{\ 8\ } \times 5 = 40$$

c. $40 \div 5 = \underline{\ 8\ }$

d. There are $\underline{\ 8\ }$ rows of plums.

UDL: Action & Expression

Consider modeling a think aloud for drawing the tape diagram from problem 2 to guide students through drawing the tape diagram for additional problems. Students may need continued support drawing a tape diagram when the number of groups is unknown. In addition, encourage students to share their thought process by asking them to think aloud and address misconceptions as needed.

UDL: Action & Expression

Consider including opportunities for students to self-reflect on their process by displaying the following sentence frames for students to refer to either independently or during partner work:

- After reading the word problem I ask myself _____.

- I look for _____.

- If I get stuck I can _____.

- It is important to _____.

Circulate and observe student work. Lead a discussion emphasizing the questions:

- What does the quotient represent: the number of groups or the size of the group? How do you know?
- How do the array and tape diagram both represent the same problem?
- Where do you see the unknown in the array? Where do you see it in the tape diagram?
- How can multiplication and division be used to solve the problem?

If time allows, direct students to problem 4 using a similar process.

Use the Read–Draw–Write process to solve the problem.

4. There are 24 students in a marching band.

    The students line up in rows of 4.

    How many rows of students are there?

 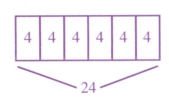

$6 \times 4 = 24$

$24 \div 4 = 6$

There are 6 rows of students.

Invite students to turn and talk about how knowing multiplication facts can help solve division problems.

Problem Set

Differentiate the set by selecting problems for students to finish independently within the timeframe. Problems are organized from simple to complex.

Teacher Note

Help students recognize the words *unknown* and *solution statement* in print on the Problem Set. Consider providing extra support, such as reading the problems aloud and underlining the terms.

Land

Debrief 5 min

Objective: Model division as an unknown factor problem.

Facilitate a discussion to emphasize the use of models to identify the unknown as the number of groups or the size of each group in multiplication and division situations.

What is the relationship between the quotient in division and the unknown factor in a related multiplication equation?

They are the same number.

They both represent the unknown in the problem.

What helps you identify the unknown as either the number of groups or the size of each group?

The words in the question of the problem help me know what I am trying to find.

When I draw a picture to represent the problem, I can see whether I need to find the number of groups or the number in each group.

Exit Ticket 5 min

Provide up to 5 minutes for students to complete the Exit Ticket. It is possible to gather formative data even if some students do not complete every problem.

Sample Solutions

Expect to see varied solution paths. Accept accurate responses, reasonable explanations, and equivalent answers for all student work.

Name _____

1. Robin puts 15 tennis balls into cans.

   She puts 3 balls in each can.

   How many cans does Robin use?

   a. Circle groups of 3 to show the balls in each can.

   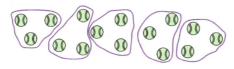

   b. Complete the equations and statements.

   $\underline{5} \times 3 = 15$

   $15 \div 3 = \underline{5}$

   Robin uses __5__ cans.

   c. What do the unknowns in the equations represent? Circle the correct answer.

   (the number of groups)   the size of each group

2. 36 apple trees are planted in rows.

   There are 4 trees in each row.

   How many rows are there?

   a. Draw an array to represent the problem.

   b. Complete the equations to find the unknown.

   $\underline{9} \times 4 = 36$

   $36 \div 4 = \underline{9}$

   c. Write a solution statement.

   There are 9 rows of apple trees.

   d. What do the unknowns in the equations represent? Circle the correct answer.

   (the number of groups)   the size of each group

3. Mrs. Smith puts a total of 12 peaches into bags.

She puts 2 peaches in each bag.

How many bags does Mrs. Smith use?

a. Draw an array to represent the problem.

b. Draw a tape diagram to represent the problem.

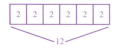

c. Complete the equations to find the unknown.

$\underline{6} \times 2 = 12$

$12 \div 2 = \underline{6}$

d. Write a solution statement.

Mrs. Smith uses 6 bags.

4. Mrs. Smith puts a total of 18 plums into bags.

She puts 6 plums in each bag.

How many bags does she use?

a. Draw an array and a tape diagram to represent the problem.

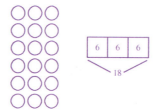

b. Complete the equations to find the unknown.

$\underline{3} \times \underline{6} = \underline{18}$

$\underline{18} \div \underline{6} = \underline{3}$

c. Write a solution statement.

Mrs. Smith uses 3 bags.

Copyright © Great Minds PBC

PROBLEM SET   129

130   PROBLEM SET

Copyright © Great Minds PBC

252

Copyright © Great Minds PBC

Model the quotient as the number of groups using units of 2, 3, 4, 5, and 10.

16

Name _____

✉ **16**

Zara uses 21 apples to make pies. She uses 3 apples for each pie. How many pies does Zara make?

a. Draw a tape diagram to represent the problem.

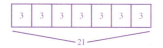

b. Write an equation to solve the problem.

$21 \div 3 = 7$

c. Zara makes ___7___ pies.

Lesson at a Glance

Students draw tape diagrams to represent division problems where the unknown, or quotient, represents the number of groups. They represent problems with division equations and unknown factor multiplication equations.

Key Questions

- How does knowing the size of the group and the total help with selecting a solution path?
- How can we use multiplication and division equations to represent the same problem?

Achievement Descriptors

3.Mod1.AD2 **Represent** a division situation with a model and **convert** between several representations of division. (3.OA.A.2)

3.Mod1.AD3 **Solve** one-step word problems by using multiplication and division within 100, involving factors and divisors 2–5 and 10. (3.OA.A.3)

3.Mod1.AD4 **Determine** the unknown number in a multiplication or division equation involving factors and divisors 2–5 and 10. (3.OA.A.4)

3.Mod1.AD7 **Represent** and **explain** division as an unknown factor problem. (3.OA.B.6)

Agenda

Fluency 10 min

Launch 5 min

Learn 35 min

- Draw to Represent Measurement Division
- Error Analysis of a Tape Diagram
- Problem Set

Land 10 min

Materials

Teacher

- None

Students

- None

Lesson Preparation

None

Fluency 🔟

Choral Response: Telling Time

Students tell time on an analog clock to the nearest half hour, using picture clues to distinguish between a.m. and p.m., to maintain work with time from grade 2.

Display the picture of the blank clock.

Let's count by 5 minutes around the clock together.

Point to the numbers on the clock as students count by fives from 0 to 60.

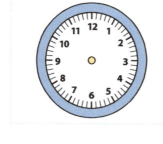

After asking each question, wait until most students raise their hands, and then signal for students to respond.

Raise your hand when you know the answer to each question.
Wait for my signal to say the answer.

Display the picture of the clock that shows 2:00.

What time does the clock show?

2:00

Is it 2:00 a.m. or p.m.?

p.m.

Continue the process with the following sequence:

| 1:00 a.m. | 7:00 a.m. | 12:30 p.m. | 8:30 a.m. | 6:30 p.m. |

Counting the Math Way by Tens and Fours

Students construct a number line while counting aloud to build fluency with counting by tens and fours and develop a strategy for multiplying.

For each skip-count, show the math way on your fingers while students count, but do not count aloud.

Let's count the math way by tens. Each finger represents 10.

Have students count the math way by tens from 0 to 100 and then back down to 0.

Hands down. Now you count aloud while I show the count on my fingers. Ready?

Lead students to count aloud forward and backward by tens, emphasize counting on from 50.

Now let's count the math way by fours. Each finger represents 4.

Have students count the math way by fours from 0 to 40 and then back down to 0.

Show me 20.

(Students show 20 on their fingers the math way.)

Have students count the math way by fours from 20 to 40 and then back down to 20.

Whiteboard Exchange: Divide Equal Groups

Students write division equations to describe an equal-groups picture to build fluency with the two interpretations of division and associated terminology.

After asking each question, wait until most students raise their hands, and then signal for students to respond.

Raise your hand when you know the answer to each question. Wait for my signal to say the answer.

Display the picture of the apples.

What is the total number of apples?

15

What is the total number of groups?

3

What is the size of each group?

5

After each prompt for a written response, give students time to work. When most students are ready, signal for students to show their whiteboards. Provide immediate and specific feedback. If students need to revise, briefly return to validate their corrections.

Write a division equation where the quotient is the size of each group.

Show the equation: $15 \div 3 = 5$

Write a division equation where the quotient is the number of groups.

Show the equation: $15 \div 5 = 3$

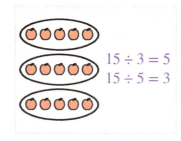

Repeat the process with the following sequence:

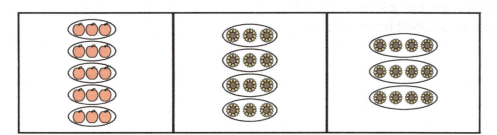

Launch ⏱ 5

Students write multiplication and division equations to represent equal groups, arrays, and tape diagrams.

Display the pictures one at a time. Ask students to write two multiplication equations and two division equations to represent each picture. Then have them box the number of groups. Students may identify the rows or columns as the number of groups for picture c and picture d.

$$\boxed{5} \times 3 = 15$$
$$3 \times \boxed{5} = 15$$
$$15 \div 3 = \boxed{5}$$
$$15 \div \boxed{5} = 3$$

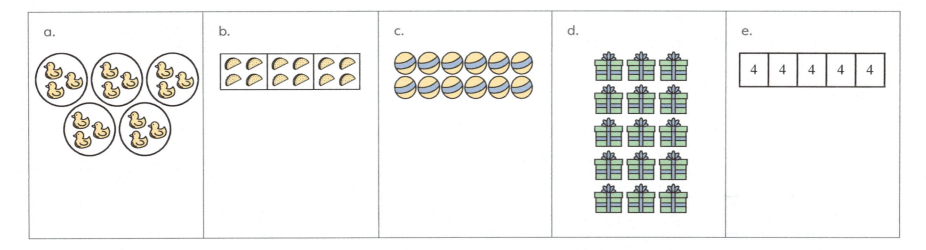

Transition to the next segment by framing the work.

Today, we will represent word problems with tape diagrams and use multiplication and division equations to find the number of groups.

Learn 35

Draw to Represent Measurement Division

Students draw a tape diagram to represent an *equal-groups with number of groups unknown* word problem and solve it with a related multiplication equation.

Direct students to problem 1 in their books. Read the entire problem chorally.

> Use the Read–Draw–Write process to solve the problem.

1. Some students share 32 sticks of sidewalk chalk.

   Each student gets 4 sticks of chalk.

   How many students get sticks of chalk?

   a. Draw a tape diagram to represent the problem.

   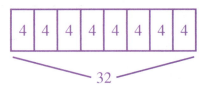

   b. Complete the two equations to represent the problem. Box the unknown.

   $\underline{\quad 32 \quad} \div \underline{\quad 4 \quad} = \boxed{8}$

   $\boxed{8} \times \underline{\quad 4 \quad} = \underline{\quad 32 \quad}$

   c. $\underline{\quad 8 \quad}$ students get sticks of chalk.

Teacher Note

Topic D provides students with their first experience using tape diagrams for division. Lessons 16 and 17 provide intentional explicit instruction on creating and interpreting different division tape diagrams. In lesson 18, students apply their understanding of tape diagrams to represent the two types of division word problems. Students may self-select their model for division in later lessons, once an understanding of using tape diagrams has been established.

Guide students to draw a tape diagram using the following sequence. Read the first sentence of the problem chorally with the class.

What do we know?

There are 32 sticks of sidewalk chalk.

We can draw a tape diagram to represent the problem. How can we show a total of 32 sticks of chalk with a tape diagram?

Draw a tape diagram, and label the whole tape 32 to represent the 32 sticks of chalk.

Read the second sentence of the problem chorally with the class.

How can we show on our tape diagram that each student gets 4 sticks of chalk?

Label a part of the tape as 4 to represent the chalk.

Draw a line to show a part of the tape diagram, and label it 4 as shown. Read the last sentence of the problem chorally with the class.

Look at our tape diagram. Is 4 the number of groups or the size of each group? How do you know?

It is the size of each group because it is how many sticks of chalk each student gets.

It is the size of the groups because I see one group of 4 in the tape diagram, and the problem did not say how many groups there are.

It is the size of each group because it is the size of one part on my tape diagram.

What division equation represents the problem?

Write $32 \div 4 =$ _____.

We can find the number of groups by asking how many fours are in 32. Let's skip-count by fours until we get to 32.

Teacher Note

In Launch, students were prompted to write two multiplication equations and two division equations to match the pictures. In Learn, students are prompted to write one equation using each operation. Without a context, multiplication and division relationships can be represented with multiple equations. However, when there is context given, there is one division equation that best represents the situation. The multiplication equation that follows the division equation supports seeing division as an unknown factor problem.

Chorally count by fours, and track the count the math way with fingers.

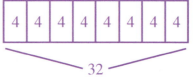

How many fours are in 32?

Ask students to write the division equation. Invite students to work with a partner to complete the tape diagram and the equation.

What do we call the 8, or the answer to a division problem?

The quotient

Complete the equation by writing 8, and draw a box around it.

What does the quotient represent in the problem?

The number of groups, or the number of students, that get chalk.

Ask students to write a multiplication equation to represent the problem and box the number of students in the equation. Prompt students to explain how both equations represent the problem with the following:

How can multiplication and division represent the same problem?

The numbers are switched around, but both multiplication and division equations show the number of groups, the size of each group, and the total.

How can we see both multiplication and division in our tape diagram?

I see multiplication because there are 8 groups of 4, and the total is 32.

I see division because the total, 32, is divided into 8 equal groups with 4 in each group.

I see the number of groups, the size of each group, and the total in the tape diagram.

Multiplication and division are related. They both show the relationship between the number of groups, the size of each group, and the total. That is why we can solve division equations by thinking about our multiplication facts.

Prompt students to complete the solution statement for the problem.

Invite students to turn and talk about specific division equations they have solved by thinking about multiplication.

Error Analysis of a Tape Diagram

Students critique a flawed response and provide a correct solution strategy.

Use the Critique a Flawed Response routine, and present the following problem:

Deepa plants 24 seeds. She plants them in rows of 3. How many rows of 3 seeds does Deepa plant?

Display the picture of the tape diagram.

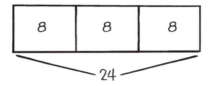

Here is a student's drawing to represent the problem, but it has an error. Can you find what the error is?

Invite students to think–pair–share to identify the error.

The problem says she planted the seeds in rows of 3. That means the size of each group is 3.

This student solved for 3 groups of 8 rather than for 8 groups of 3.

Give students a few minutes to represent the problem with a tape diagram and solve the problem. A sample tape diagram and solution are shown here. Circulate and identify a few students to share their thinking.

| 3 | 3 | 3 | 3 | 3 | 3 | 3 | 3 |

$\boxed{8} \times 3 = 24$ Deepa plants 8 rows of 3.

$24 \div 3 = \boxed{8}$

24

Then facilitate a class discussion. Invite students to share their solutions with the class. Lead the class to consensus about how to correct the flawed response.

As time permits, direct students to solve problem 2 and problem 3. Circulate as students work, and guide them to draw a tape diagram to represent the problem.

Language Support

Encourage the use of the Talking Tool to help students engage in the conversation. The tool will help students consider what questions to ask and will initiate student-to-student discourse.

Promoting the Standards for Mathematical Practice

When students think–pair–share about a flawed "mystery student's" work, they are constructing viable arguments and critiquing the reasoning of others (MP3).

Ask the following questions to promote MP3:

- What parts of the student's drawing do you question? Why?

- How would you change the student's drawing to make it more accurate?

Use the Read–Draw–Write process to solve the problem.

2. A pet store has 14 birds.

   There are 2 birds in each cage.

   How many cages of birds are there?

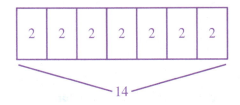

$\boxed{7} \times 2 = 14$

$14 \div 2 = \boxed{7}$

There are 7 cages of birds.

Use the Read–Draw–Write process to solve the problem.

3. A classroom has enough tables to seat a total of 20 students.

   There are 4 students seated at each table.

   How many tables are in the classroom?

$\boxed{5} \times 4 = 20$

$20 \div 4 = \boxed{5}$

There are 5 tables in the classroom.

Invite students to turn and talk about how to draw tape diagrams when the total and the size of each group are known and the number of groups is unknown.

Problem Set

Differentiate the set by selecting problems for students to finish independently within the timeframe. Problems are organized from simple to complex.

UDL: Representation

Consider presenting several word problem scenarios and asking students if the problem tells the size of each group or the number of groups. Have students show 1 finger if the size of the groups is known and 2 fingers if the number of groups is known. For example:

• Some friends share 21 markers. Each student gets 3 markers. How many students get markers? What else do we know besides the total, the size of the groups (show 1 finger), or the number of groups (show 2 fingers)?

• Zara has 24 stickers. 6 friends share the stickers. How many stickers does each friend get? What else do we know besides the total, the size of the groups (show 1 finger), or the number of groups (show 2 fingers)?

Continue with examples as needed to ensure students can differentiate between the number of groups and the size of the groups in the context of word problems.

Land 10

Debrief 5 min

Objective: Model the quotient as the number of groups using units of 2, 3, 4, 5, and 10.

Facilitate a discussion emphasizing the relationship between multiplication and division and the use of tape diagrams to represent division problems when the number of groups is unknown.

Display the picture of the tape diagram.

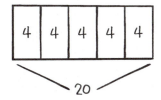

How does knowing the size of the group and the total help with selecting a solution path?

If I know the total and the size of each group, then I know I need to find the number of groups. That tells me I can divide or think of it as an unknown factor problem.

How can we use multiplication and division equations to represent the same problem?

Both multiplication and division equations show the same three things—the total, the number of groups, and the size of the groups.

Multiplication and division are related like addition and subtraction are related.

Exit Ticket 5 min

Provide up to 5 minutes for students to complete the Exit Ticket. It is possible to gather formative data even if some students do not complete every problem.

Sample Solutions

Expect to see varied solution paths. Accept accurate responses, reasonable explanations, and equivalent answers for all student work.

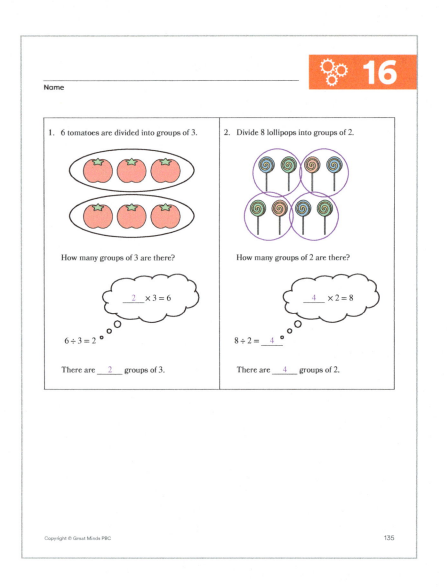

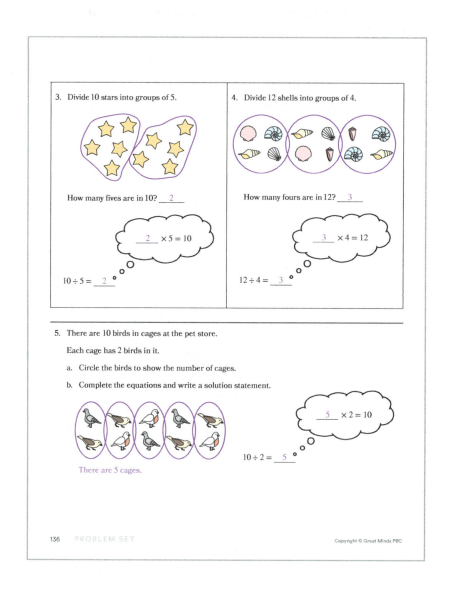

Use the Read–Draw–Write process to solve each problem.

6. Adam buys 24 meters of wire.

   He cuts the wire into pieces that are each 4 meters long.

   How many pieces of wire does he cut?

   $24 \div 4 = 6$

   Adam cuts 6 pieces of wire.

7. Eva makes 24 pancakes and puts them into stacks.

   There are 6 pancakes in each stack.

   How many stacks of pancakes are there?

   $24 \div 6 = 4$

   There are 4 stacks of pancakes.

Model the quotient as the size of each group using units of 2, 3, 4, 5, and 10.

✉ **17**

Name _____

Luke has 14 oat bars.

He eats the same number of oat bars each day. He eats all the oat bars in 7 days.

How many oat bars does Luke eat each day?

a. Draw a tape diagram to represent the problem.

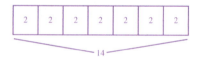

b. Write an equation to solve the problem.
    $14 \div 7 = 2$

c. Luke eats ___2___ oat bars each day.

Lesson at a Glance

Students draw tape diagrams to represent division problems where the group size is unknown. They represent problems with division equations and unknown factor multiplication equations.

Key Questions

- How does the tape diagram help us see relationships and select a solution path?

- How can we use multiplication and division equations to represent the same problem?

Achievement Descriptors

3.Mod1.AD2 Represent a division situation with a model and **convert** between several representations of division. (3.OA.A.2)

3.Mod1.AD3 Solve one-step word problems by using multiplication and division within 100, involving factors and divisors 2–5 and 10. (3.OA.A.3)

3.Mod1.AD4 Determine the unknown number in a multiplication or division equation involving factors and divisors 2–5 and 10. (3.OA.A.4)

3.Mod1.AD7 Represent and **explain** division as an unknown factor problem. (3.OA.B.6)

Agenda

Fluency 10 min

Launch 5 min

Learn 35 min

- Draw to Represent Partitive Division
- Partitive Division Word Problems
- Problem Set

Land 10 min

Materials

Teacher

- None

Students

- None

Lesson Preparation

None

Fluency

Choral Response: Telling Time

Students tell time on an analog clock to the nearest five minutes, using picture clues to distinguish between a.m. and p.m., to maintain work with time from grade 2.

After asking each question, wait until most students raise their hands, and then signal for students to respond.

Raise your hand when you know the answer to each question. Wait for my signal to say the answer.

Display the picture of the clock and the child in bed.

What time does the clock show?

7:15

Is it 7:15 a.m. or p.m.?

a.m.

Repeat the process with the following sequence:

Counting the Math Way by Fives and Threes

Students construct a number line while counting aloud to build fluency with counting by fives and threes and develop a strategy for multiplying.

For each skip-count, show the math way on your fingers while students count, but do not count aloud.

Let's count the math way by fives. Each finger represents 5.

Have students count the math way by fives from 0 to 50 and then back down to 0.

Hands down. Now you count aloud while I show the count on my fingers. Ready?

Lead students to count aloud forward and backward by fives and emphasize counting on from 25.

Now let's count the math way by threes. Each finger represents 3.

Have students count the math way by threes from 0 to 30 and then back down to 0.

Show me 15.

(Students show 15 on their fingers the math way.)

Have students count the math way by threes from 15 to 30 and then back down to 15.

Hands down. Now you count aloud while I show the count on my fingers. Ready?

Lead students to count forward and backward by threes, and emphasize counting on from 15.

Whiteboard Exchange: Divide Equal Groups

Students write division equations to describe an equal-groups picture to build fluency with two interpretations of division and associated terminology.

After asking each question, wait until most students raise their hands, and then signal for students to respond.

Raise your hand when you know the answer to each question. Wait for my signal to say the answer.

Display the picture of the apples.

What is the total number of apples?

8

What is the total number of groups?

2

What is the size of each group?

4

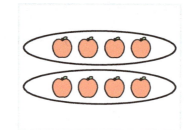

After each prompt for a written response, give students time to work. When most students are ready, signal for students to show their whiteboards. Provide immediate and specific feedback. If students need to revise, briefly return to validate their corrections.

Write a division equation where the quotient is the size of each group.

Show the equation: $8 \div 2 = 4$

Write a division equation where the quotient is the number of groups.

Show the equation: $8 \div 4 = 2$

Repeat the process with the following sequence:

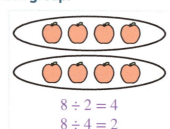

$8 \div 2 = 4$
$8 \div 4 = 2$

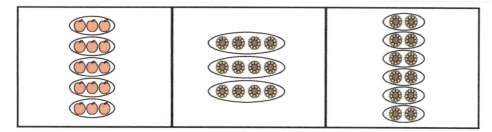

 **Launch** ⏱ 5

Students write multiplication and division equations to represent a picture.

Display the pictures one at a time. Ask students to write two multiplication equations and two division equations to represent each picture. Then have them box the size of the group in each equation. Students may identify the rows or columns as the size of the group for picture c and picture d.

$$5 \times \boxed{3} = 15$$
$$\boxed{3} \times 5 = 15$$
$$15 \div \boxed{3} = 5$$
$$15 \div 5 = \boxed{3}$$

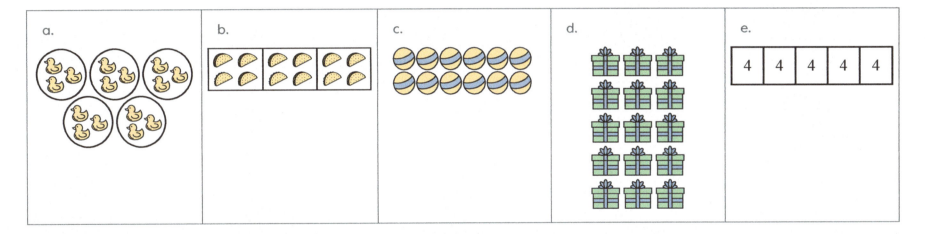

Transition to the next segment by framing the work.

Today, we will represent word problems with tape diagrams and use multiplication and division equations to find the size of each group.

Learn 35

Draw to Represent Partitive Division

Students draw a tape diagram to represent an *equal-groups with group size unknown* **word problem and solve it with a related multiplication equation.**

Direct students to problem 1 in their books. Read the entire problem chorally.

Use the Read–Draw–Write process to solve the problem.

1. Three students share 27 crackers equally.

   How many crackers does each student get?

   a. Draw a tape diagram to represent the problem.

   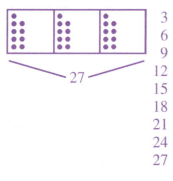

   3
   6
   9
   12
   15
   18
   21
   24
   27

   b. Complete a division equation and a multiplication equation to represent the problem. Box the unknown.

   $$\underline{27} \div \underline{3} = \boxed{9}$$

   $$\underline{3} \times \boxed{9} = \underline{27}$$

   c. Each student gets __9__ crackers.

Guide students to draw a tape diagram using the following sequence. Read the first sentence of the problem chorally with the class.

What do we know?

3 students equally share 27 crackers.

We can draw a tape diagram to represent the problem. Do we know the total?

Yes.

What is the total?

27 crackers

How can we show 27 crackers on our tape diagram?

Label the whole tape 27 to represent the 27 crackers.

Draw a tape diagram and label 27 as the total.

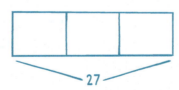

What happens with the 27 crackers?

The 27 crackers are equally shared by 3 students.

Is 3 the number of groups or the size of each group?

Invite students to turn and talk to verify that 3 students is the number of groups.

How can we show this on our tape diagram?

We can partition the tape diagram into 3 equal parts.

Partition the tape diagram into 3 equal parts.

Read the last sentence of the problem chorally with the class.

What is the unknown?

The number of crackers each student gets.

What division equation represents the problem?

Write $27 \div 3 = $ _____.

Look at your tape diagram. Does it represent everything the problem told us? How do you know?

Yes, the tape diagram shows there are 27 crackers and 3 groups.

Watch me equally share the crackers in a way that is different from what we have done before.

Model equally sharing the crackers on the tape diagram. Give each student 1 cracker. Track the running total of the number shared with a skip-count, along the side of the tape diagram.

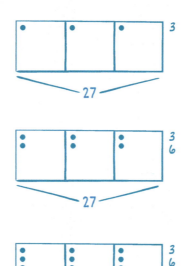

I gave each student 1 cracker. How many total crackers did I give out?

Do I have enough to give them all 1 more?

Draw another dot in each of the parts of the tape diagram to equally share 3 more crackers.

How many total crackers did I give out?

I'll keep track of the total crackers with a skip-count, like this.

Continue to equally share the crackers and record the running total next to the tape diagram. When finished ask:

How many crackers does each student get? How do you know?

Each student gets 9 crackers. I counted the dots in one group of the tape diagram.

They each get 9 crackers because I counted how many times you skip-counted by threes. There are 9 threes in 27.

Direct students to work with a partner to solve as modeled and complete the tape diagram and skip-count.

What do we call 9, or the answer, to a division problem?

The quotient

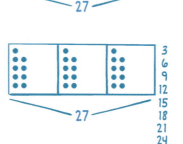

Teacher Note

By modeling the skip-count along with the equal sharing, students see skip-counting as a strategy for both interpretations of division, partitive and measurement. Students can solve this problem by skip-counting because they are giving out a total of 3 crackers at a time, as they give 1 cracker to each student.

As students become more comfortable with using skip-counting as a solution strategy, they will no longer need to equal share within the tape diagram with dots and can move to a more abstract representation. Students may write the size of the group in each part of the tape diagram instead.

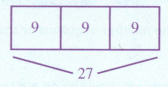

Draw a box around 9.

What does the quotient represent in the problem?

The quotient represents the size of each group, or the number of crackers each student gets.

Write a multiplication equation that represents the problem. Box the size of each group.

Write $27 \div 3 = 9$. Invite students to turn and talk about how the equation represents the problem.

Prompt students to complete a solution statement.

Invite students to think–pair–share about what makes the tape diagram a useful representation.

The tape diagram is just like our equal-groups model. They both help me see the number of groups, the size of the groups, and the total.

I can represent each part of the problem with the tape diagram. It helps me to know how to represent the problem with an equation.

Invite students to think–pair–share about how this tape diagram is different from tape diagrams that have been used in previous lessons.

Today we showed the number of groups on a tape diagram. Last time we showed the size of each group on a tape diagram.

When the number of groups is unknown, we draw one part at a time. When the size of each group is unknown, we partition the tape into the number of groups and equally share the total.

Language Support

Show the completed tape diagram, and prompt students to connect the pictorial representation to the strategy.

- Point to where the drawing shows the number of groups. This is the number of students who share the crackers.

- Point to where the drawing shows the size of each group. This is the number of crackers each student gets.

- Show me how the drawing shows sharing the crackers as "one for you, one for you."

Partitive Division Word Problems

Students draw tape diagrams to represent and solve *equal-groups with group size unknown* word problems.

Direct students to problem 2. Ask them to use the Read–Draw–Write process and work with a partner to solve the problem. Encourage students to draw a tape diagram to represent the problem.

> Use the Read–Draw–Write process to solve the problem.
>
> 2. Mr. Endo puts 28 papers equally into 4 piles.
>
>   How many papers are in each pile?
>
>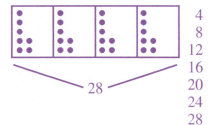
>
>   4
>   8
>   12
>   16
>   20
>   24
>   28
>
> $28 \div 4 = \boxed{7}$
>
> $4 \times \boxed{7} = 28$
>
>   There are 7 papers in each pile.

Circulate and observe student work. Select a few students who correctly draw a tape diagram to share their work.

Consider using the following suggested questions to guide student sharing:

- What do we know? What are we looking for?
- How did you represent the problem with a tape diagram?
- How does your tape diagram help you write an equation and select a strategy to solve?
- What division equation represents the problem?
- What unknown factor multiplication equation represents the problem?
- What strategy did you use to solve? Why?
- What does the quotient represent?

Promoting the Standards for Mathematical Practice

When students draw tape diagrams and write equations to represent the problem, they are modeling with mathematics (MP4). In particular, students analyze the efficiency of their models and consider when certain models are more appropriate in certain contexts. Ask the following questions to promote MP4:

- What key ideas in problem 2 do you need to make sure are in your model?

- What does each number in your equation tell you about the papers?

Teacher Note

Accept any accurate representation, including equal sharing within the tape diagram with dots or the more abstract tape as shown.

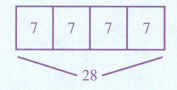

The drawing should make sense to the student using it and correctly represent the problem. The student should be able to accurately explain how the tape diagram relates to the problem.

If time allows, direct students to problem 3 using a similar procedure.

Use the Read–Draw–Write process to solve the problem.

3. Shen puts a total of 45 marbles equally into 5 bags.

How many marbles are in each bag?

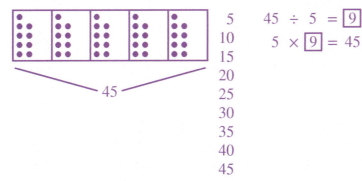

$$45 \div 5 = \boxed{9}$$
$$5 \times \boxed{9} = 45$$

5
10
15
20
25
30
35
40
45

There are 9 marbles in each bag.

Invite students to turn and talk about how they used skip-counting to track their equal shares.

UDL: Action & Expression

To support the use of a tape diagram for this interpretation of division, consider modeling drawing one group at a time instead of drawing the entire tape diagram and then partitioning it into equal groups. Inch grid paper may help with creating same-sized groups.

Problem Set

Differentiate the set by selecting problems for students to finish independently within the timeframe. Problems are organized from simple to complex.

Debrief 5 min

Objective: Model the quotient as the size of each group using units of 2, 3, 4, 5, and 10.

Facilitate a discussion that emphasizes how the tape diagram represents division problems when group size is unknown.

How are the tape diagrams we have been using today helpful in selecting a solution path?

Tape diagrams help me see the total and how many groups there are so that I can find the group size. That tells me I can divide or think of it as an unknown factor problem.

How can we use multiplication and division equations to represent the same problem?

Both multiplication and division equations show the same three things—the total, the number of groups, and the size of groups.

Multiplication and division are related like addition and subtraction are related.

Exit Ticket 5 min

Provide up to 5 minutes for students to complete the Exit Ticket. It is possible to gather formative data even if some students do not complete every problem.

Sample Solutions

Expect to see varied solution paths. Accept accurate responses, reasonable explanations, and equivalent answers for all student work.

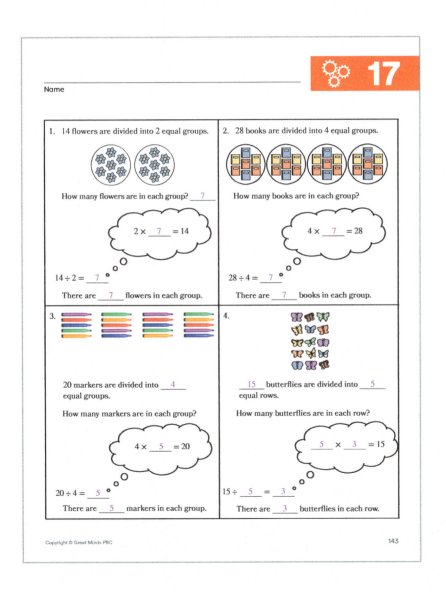

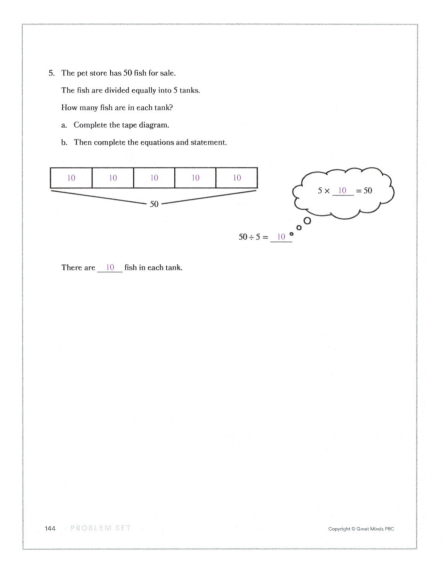

6. Mr. Davis has 24 pencils.

He divides them equally among 4 tables.

a. How many pencils are on each table?

b. Complete the tape diagram. Then complete the equations and statement.

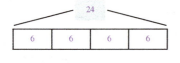

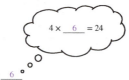

$4 \times \underline{\ 6\ } = 24$

$24 \div \underline{\ 4\ } = \underline{\ 6\ }$

There are __6__ pencils on each table.

Use the Read–Draw–Write process to solve the problem.

7. 24 jelly beans are shared equally by 3 students.

How many jelly beans does each student get?

$24 \div 3 = 8$

Each student gets 8 jelly beans.

18

Represent and solve measurement and partitive division word problems.

Name _____

✉ 18

Mia puts 20 books into 4 equal piles.

How many books are in each pile?

a. Draw a picture to represent the problem.

Sample:

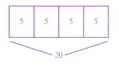

b. Write an equation to represent the problem.

20 ÷ 4 = 5

c. There are ___5___ books in each pile.

Lesson at a Glance

Students solve two types of division word problems—measurement and partitive—and examine a solution strategy for each problem. A video provides context for the word problems. Students answer questions to frame their thinking about how the context provides support for representing each type of problem.

Key Questions

- What is the difference between the two types of division word problems?
- How can we use tape diagrams to represent the two types of division word problems?

Achievement Descriptors

3.Mod1.AD2 **Represent** a division situation with a model and **convert** between several representations of division. (3.OA.A.2)

3.Mod1.AD3 **Solve** one-step word problems by using multiplication and division within 100, involving factors and divisors 2–5 and 10. (3.OA.A.3)

Agenda

Fluency 15 min

Launch 10 min

Learn 25 min

- Compare Measurement and Partitive Division Word Problems
- Apply the RDW Process to Two Types of Division Word Problems
- Problem Set

Land 10 min

Materials

Teacher

- None

Students

- Count by Tens and Fives Sprint (in the student book)
- Tape Diagram and Equation Card Sort (1 per student pair, in the student book)
- Scissors (1 per student pair)

Lesson Preparation

- Consider tearing out the Sprint pages in advance of the lesson.
- Tear out and cut apart the Tape Diagram and Equation Card Sort cards from the student book. One set of cards per student pair is needed. Consider whether to prepare these materials in advance or have students assemble them during the lesson.

Sprint: Count by Tens and Fives

Materials—S: Count by Tens and Fives Sprint

Students write the unknown number in a sequence to build fluency with counting by tens and fives.

Have students read the instructions and complete the sample problems.

Fill in the blank to complete the sequence.

1.	10, 20, 30, _____	40
2.	45, 40, 35, _____	30

Direct students to Sprint A. Frame the task.

I do not expect you to finish. Do as many problems as you can, your personal best.

Take your mark. Get set. Think!

Time students for 1 minute on Sprint A.

Stop! Underline the last problem you did.

I'm going to read the answers. As I read the answers, call out "Yes!" if you got it correct. If you made a mistake, circle the answer.

Read the answers to Sprint A quickly and energetically.

Count the number you got correct and write the number at the top of the page. This is your personal goal for Sprint B.

Celebrate students' effort and success.

Provide about 2 minutes to allow students to complete more problems or to analyze and discuss patterns in Sprint A. If students are provided time to complete more problems on Sprint A, reread the answers but do not have them alter their personal goals.

Lead students in one fast-paced and one slow-paced counting activity, each with a stretch or physical movement.

Point to the number you got correct on Sprint A. Remember this is your personal goal for Sprint B.

Direct students to Sprint B.

Take your mark. Get set. Improve!

Time students for 1 minute on Sprint B.

Stop! Underline the last problem you did.

I'm going to read the answers. As I read the answers, call out "Yes!" if you got it correct. If you made a mistake, circle the answer.

Read the answers to Sprint B quickly and energetically.

Count the number you got correct and write the number at the top of the page.

Figure out your improvement score and write the number at the top of the page.

Celebrate students' improvement.

> **Teacher Note**
>
> Consider asking the following questions to discuss the patterns in Sprint A:
>
> - How do problems 1–6 compare to problems 7–12?
>
> - What do you notice about problems 13–22?

> **Teacher Note**
>
> Count forward by fives from 0 to 50 for the fast-paced counting activity.
>
> Count backward by fives from 50 to 0 for the slow-paced counting activity.

Launch 10

Materials—S: Scissors, Tape Diagram and Equation Card Sort

Students match word problems with tape diagrams and equations, focusing on how the information in a word problem provides support with how to start drawing a tape diagram.

Pair students. Direct students to the Tape Diagram and Equation Card Sort. Direct one student in each pair to tear out the page, and direct partners to cut out the cards. Then direct partners to match each problem with the unfinished tape diagram and equation that represents the problem.

Provide partners time to work. As partners work, consider asking the following questions:

- How do you know this is a division problem?
- In the story, we know the total. What else do we know in the story—the number of groups or the size of each group?
- Look at the tape diagram. Do you know the number of groups or the size of each group?
- What is the unknown?
- What does the number 10 (or 6) represent in the problem?

Gather the class with their matched cards for discussion. Verify that partners have accurate matches.

Differentiation: Support

Instead of sorting the problems, tape diagrams, and equations simultaneously, students may find it easier to first match the problems and equations, and then add the tape diagrams. The numbers in the equation match the numbers in the word problem regardless of the interpretation of the division.

Matching the word problems to the tape diagrams involves identifying whether the division is partitive or measurement. Keep the focus on the known and unknown information in the problem and what is known and unknown in the tape diagrams to help students reason about the correct matches.

There are 60 cookies packed equally in boxes. Each box has 10 cookies. How many boxes of cookies are there?		$60 \div 10 = 6$
There are 60 cookies packed equally in boxes. There are 6 boxes. How many cookies are in each box?		$60 \div 6 = 10$
There are 60 cookies packed equally in boxes. There are 10 boxes. How many cookies are in each box?		$60 \div 10 = 6$
There are 60 cookies packed equally in boxes. Each box has 6 cookies. How many boxes of cookies are there?		$60 \div 6 = 10$

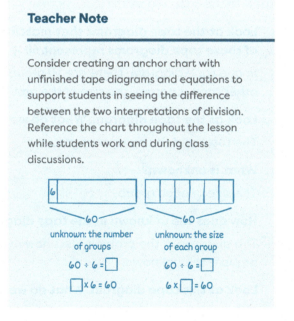

Teacher Note

Consider creating an anchor chart with unfinished tape diagrams and equations to support students in seeing the difference between the two interpretations of division. Reference the chart throughout the lesson while students work and during class discussions.

Facilitate a conversation about the similarities and differences in the problems, tape diagrams, and equations. The goal is for students to see that:

- Sometimes the unknown is the number of groups and sometimes it is the size of the groups.

- The same equation can describe two different division situations.

Consider asking the following questions:

- How are the problems, tape diagrams, and equations similar? How are they different?

- Where do we see the total, number of groups, and size of each group?

Refer to the tape diagrams that represent the equation $60 \div 10 = 6$.

Look at the tape diagrams that match the equation $60 \div 10 = 6$. How do each of these tape diagrams represent $60 \div 10 = 6$?

In one tape diagram, there are 10 groups, so each group must have a size of 6. In the other tape diagram, the size of the group is 10, so there must be 6 groups.

Look at the tape diagram. What do we know?

The total and the number of groups.

What is unknown?

The size of each group.

How does the unknown in the tape diagram relate to the problem?

The question in the problem asks how many cookies are in each box, so the size of each group is the unknown.

Look at this tape diagram. What do we know?

The total and the size of each group.

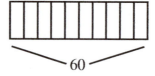

What is unknown?

The number of groups

How does the unknown in the tape diagram relate to the problem?

The question in the problem asked how many boxes there are, so the number of groups is unknown.

Ask similar questions to compare the tape diagrams that represent the equation $60 \div 6 = 10$.

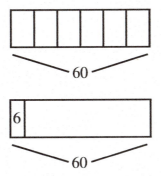

Guide students to the conclusion that the unknown in the context determines how the problem is represented with a drawing. Sometimes different drawings can be represented by the same equation.

Transition to the next segment by framing the work.

Today, we will solve two types of division word problems. Sometimes the number of groups will be unknown, and other times the size of the groups will be unknown.

 **25**

Compare Measurement and Partitive Division Word Problems

Students collect information from a video, solve division word problems, and reason about the similarities and differences between measurement and partitive division.

Play part 1 of the Boxing Cookies video. If necessary, replay the video and ask students to note any details.

Give students 1 minute to turn and talk about what they noticed.

Engage students in a brief conversation about the video. Consider the following possible sequence:

What do you notice?

She baked 24 cookies.

She put 6 cookies in a box.

What do you wonder?

How many boxes will she fill with cookies?

Let's keep watching. As you watch, think about what changes.

Play part 2 of the Boxing Cookies video. If necessary, replay the video and ask students to note any details.

Give students 1 minute to turn and talk about what they noticed.

Engage students in a brief conversation about the video. Listen for students to make connections between the two parts of the video. Consider the following possible sequence:

What do you notice?

She baked 24 cookies again.

She has 6 boxes. The boxes are smaller.

She hasn't put any cookies in a smaller box yet.

What do you wonder?

How many cookies will she put in each smaller box?

There are many mathematical questions we could ask. Let's use what we saw in the video to help us understand and solve two word problems.

Direct students to problem 1 in their books. Read the entire problem chorally.

Use the Read–Draw–Write process to solve the problem.

1. There are 24 cookies packed in boxes.

   Each box has 6 cookies.

How many boxes are needed to package all the cookies?

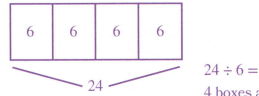

_____ × 6 = 24

6, 12, 18, 24

24 ÷ 6 = 4

4 boxes are needed.

Use the Read–Draw–Write process to solve the problem.

2. There are 24 cookies packed equally in 6 boxes.

   How many cookies are packed in each box?

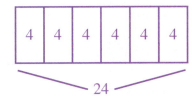

24 ÷ 6 = 4

There are 4 cookies in each box.

Prompt students to reason about the situation by rereading the problem and asking questions such as:

- What do you know?
- What are you trying to find?
- How could a tape diagram represent the problem?

Have students turn and talk about how to represent the problem with a tape diagram.

Use a similar sequence to facilitate a discussion about the context and tape diagram for problem 2.

Direct students to work with a partner to draw a tape diagram, write an equation, and write a solution statement for each problem.

Provide partners time to work. As partners work, circulate and observe the tape diagrams and equations they use to represent the problems.

Promoting the Standards for Mathematical Practice

As students extract the relevant information from the word problem, write and evaluate expressions, then relate their solution back to the original context, they are reasoning quantitatively and abstractly (MP2).

Ask the following questions to promote MP2:

- What does the given information in each problem tell you about drawing your tape diagram?

- How do your tape diagrams represent the problems?

- Does your answer make sense in the original problem?

UDL: Action & Expression

Consider providing students one of each type of tape diagram card from Launch to use as a model for creating tape diagrams for the new problem. Invite partners to select the drawing that represents the same action and use it to create a tape diagram for the new problem. Offering this type of scaffold supports students as they develop fluency with creating tape diagrams.

Invite partners to share their drawings and strategies by asking questions such as:

How are the problems similar?

Both problems have 24 cookies that are put in boxes.

How are the problems different?

Problem 1 asks how many boxes are needed and problem 2 asks how many cookies are in each box.

What is unknown in problem 1—the number of groups or the size of each group? In problem 2?

In problem 1, the number of groups is unknown. In problem 2, the size of the group is unknown.

Explain how you started drawing your tape diagram for problem 1.

Since we know the total is 24, we drew a tape diagram and labeled the total tape 24. The size of each group is 6, so we drew just one part and labeled it 6.

Explain how you started drawing your tape diagram for problem 2.

Since we know there are 6 boxes, we drew a tape diagram with 6 parts.

Invite students to turn and talk about why the equations for both problems are the same if the situations are different.

Apply the RDW Process to Two Types of Division Word Problems

Students reason about, represent, and solve an *equal groups with group size unknown* word problem and an *equal groups with number of groups unknown* word problem.

Direct partners to problem 3 and problem 4. Instruct students to draw a tape diagram to represent the problems.

> Use the Read–Draw–Write process to solve the problem.
>
> 3. Jayla has 20 toy cars.
>
>   She puts the cars in 4 equal groups.

How many cars are in each group?

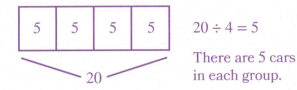

$20 \div 4 = 5$

There are 5 cars in each group.

Use the Read–Draw–Write process to solve the problem.

4. The students on the team equally share 12 apples.

   Each student gets 2 apples.

   How many students are on the team?

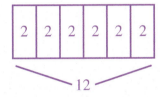

$12 \div 2 = 6$ 2, 4, 6, 8, 10, 12

There are 6 students on the team.

Provide partners time to work. As partners work, circulate and observe the tape diagrams used to represent the problems. Select partners to share their work with the class.

Invite partners to share their drawings and strategies by asking questions such as:

• What is unknown in problem 3—the number of groups or the size of each group? In problem 4?

• How did the unknown help you know how to draw the tape diagram?

• How did the tape diagram help you know what equation to write?

Problem Set

Differentiate the set by selecting problems for students to finish independently within the timeframe. Problems are organized from simple to complex.

Land

Debrief 5 min

Objective: Represent and solve measurement and partitive division word problems.

Facilitate a discussion emphasizing the differences between partitive and measurement division.

What is the difference between the two types of division word problems?

Sometimes we know the total and the number of groups, and sometimes we know the total and the size of the group.

In one type of division word problem, we are trying to find the size of each group. In the other type, we are trying to find the number of groups. Why is understanding what is unknown useful when solving problems?

It helps us know if we need to find the number of groups or the size of each group.

It can help us to make sense of the problem and make a plan to solve the problem.

How can we use tape diagrams to represent the two types of division word problems?

When we know the total and the number of groups, we can make a tape diagram with the number of parts to match the number of groups.

When we know the total and the size of each group, we make a tape diagram with an unknown number of parts. We draw one part and label it with the group size.

Exit Ticket 5 min

Provide up to 5 minutes for students to complete the Exit Ticket. It is possible to gather formative data even if some students do not complete every problem.

Sample Solutions

Expect to see varied solution paths. Accept accurate responses, reasonable explanations, and equivalent answers for all student work.

A

Number Correct: _____

Fill in the blank to complete the sequence.

1.	0, 10, 20, _____	30	23.	30, 40, _____, 60	50
2.	50, 60, 70, _____	80	24.	_____, 70, 80, 90	60
3.	70, 80, 90, _____	100	25.	_____, 30, 40, 50	20
4.	30, 20, 10, _____	0	26.	_____, 40, 30, 20	50
5.	70, 60, 50, _____	40	27.	_____, 50, 40, 30	60
6.	100, 90, 80, _____	70	28.	_____, 90, 80, 70	100
7.	0, 5, 10, _____	15	29.	25, 30, _____, 40	35
8.	25, 30, 35, _____	40	30.	_____, 30, 35, 40	25
9.	35, 40, 45, _____	50	31.	_____, 15, 20, 25	10
10.	15, 10, 5, _____	0	32.	_____, 35, 30, 25	40
11.	40, 35, 30, _____	25	33.	_____, 45, 40, 35	50
12.	50, 45, 40, _____	35	34.	_____, 20, 15, 10	25
13.	20, 30, _____, 50	40	35.	90, 100, 110, _____	120
14.	70, _____, 90, 100	80	36.	140, 130, _____, 110	120
15.	50, 40, _____, 20	30	37.	60, 65, 70, _____	75
16.	30, _____, 10, 0	20	38.	85, 80, _____, 70	75
17.	100, _____, 80, 70	90	39.	180, _____, 200, 210	190
18.	0, 5, _____, 15	10	40.	_____, 200, 190, 180	210
19.	35, _____, 45, 50	40	41.	95, _____, 105, 110	100
20.	15, 10, _____, 0	5	42.	_____, 100, 95, 90	105
21.	45, _____, 35, 30	40	43.	_____, 170, 180	160
22.	50, _____, 40, 35	45	44.	_____, 105, 100	110

B

Number Correct: _____

Improvement: _____

Fill in the blank to complete the sequence.

1.	10, 20, 30, _____	40	23.	20, 30, _____, 50	40
2.	40, 50, 60, _____	70	24.	_____, 60, 70, 80	50
3.	70, 80, 90, _____	100	25.	_____, 20, 30, 40	10
4.	40, 30, 20, _____	10	26.	_____, 30, 20, 10	40
5.	60, 50, 40, _____	30	27.	_____, 40, 30, 20	50
6.	100, 90, 80, _____	70	28.	_____, 90, 80, 70	100
7.	5, 10, 15, _____	20	29.	15, 20, _____, 30	25
8.	20, 25, 30, _____	35	30.	_____, 20, 25, 30	15
9.	35, 40, 45, _____	50	31.	_____, 10, 15, 20	5
10.	20, 15, 10, _____	5	32.	_____, 25, 20, 15	30
11.	35, 30, 25, _____	20	33.	_____, 45, 40, 35	50
12.	50, 45, 40, _____	35	34.	_____, 15, 10, 5	20
13.	10, 20, _____, 40	30	35.	80, 90, 100, _____	110
14.	60, _____, 80, 90	70	36.	130, 120, _____, 100	110
15.	40, 30, _____, 10	20	37.	50, 55, 60, _____	65
16.	40, _____, 20, 10	30	38.	75, 70, _____, 60	65
17.	100, _____, 80, 70	90	39.	170, _____, 190, 200	180
18.	5, 10, _____, 20	15	40.	_____, 190, 180, 170	200
19.	30, _____, 40, 45	35	41.	85, _____, 95, 100	90
20.	20, 15, _____, 5	10	42.	_____, 95, 90, 85	100
21.	35, _____, 25, 20	30	43.	_____, 160, 170	150
22.	50, _____, 40, 35	45	44.	_____, 100, 95	105

18

Name _____

1. Miss Diaz puts 12 frogs into 4 equal groups.

   How many frogs are in each group?

   a. Circle the tape diagram that represents the problem.

   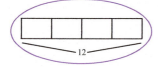 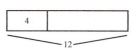

   b. Complete the equation and statement.

   $12 \div 4 =$ ___3___

   There are ___3___ frogs in each group.

2. Luke puts 21 cups of water equally into some bottles.

   He puts 3 cups of water into each bottle.

   How many bottles does Luke put water into?

   a. Circle the tape diagram that represents the problem.

   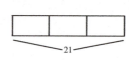 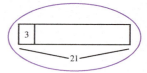

   b. Complete the equation and statement.

   $21 \div 3 =$ ___7___

   Luke puts water into ___7___ bottles.

Use the Read–Draw–Write process to solve each problem.

3. A baker packs 70 muffins in 10 boxes.

   How many muffins does the baker pack in each box?

   $70 \div 10 = 7$

   The baker packs 7 muffins in each box.

4. The server puts 35 glasses into rows of 5.

   How many rows does the server make?

   $35 \div 5 = 7$

   The server makes 7 rows of glasses.

5. Shen pays $27 for some notebooks.

   Each notebook costs $3.

   How many notebooks does Shen buy?

   $27 \div 3 = 9$

   Shen buys 9 notebooks.

6. Pablo and Casey buy 2 tickets to the movies.

   The tickets cost a total of $16.

   Pablo and Casey share the cost equally.

   How much does Casey pay?

   $16 \div 2 = 8$

   Casey pays $8.

Topic E
Application of Multiplication and Division Concepts

In topic E, students apply what they learned in previous topics to more complex problems, including those with larger factors. Through exploration, students become more flexible with using the properties of multiplication. They represent facts in ways that make sense to them and enhance their fluency with multiplication and division.

The break apart and distribute strategy is used to help students multiply and divide. Emphasis is placed on breaking apart arrays into fives facts and another fact when a factor is larger than 5. Other efficient ways to break apart arrays, such as forming doubles, are discussed as well.

Equal groups are revisited in topic E to lay a foundation for the associative property of multiplication. Students break an array into equal groups and describe the resulting groups using a multiplication expression with three factors.

In the final two lessons of the topic, students solve two-step word problems involving all four operations. They represent the problems using drawings such as arrays, number bonds, and tape diagrams, and they select appropriate solution strategies based on their understanding of the problem situations. Class discussions follow. Explaining, justifying, and comparing their strategies with peers improve students' ability to make sense of future problems and approach them with confidence.

In module 2, students build fluency with multiplication and division facts through fluency activities. In module 3, students use the associative and distributive properties as strategies for working with the remaining single-digit factors (i.e., 6, 7, 8, 9, 0, and 1) and factors that are two-digit multiples of 10. Flexibility in the use of the properties fosters a deep understanding that supports students with the more abstract property work and formal study of algebraic concepts that are introduced in later years.

Progression of Lessons

Lesson 19

Use the distributive property to break apart multiplication problems into known facts.

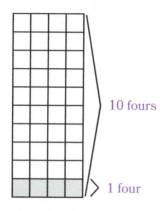

10 fours

1 four

$9 \text{ fours} = 10 \text{ fours} - 1 \text{ four}$

$9 \times 4 = (10 \times 4) - (1 \times 4)$

$9 \times 4 = 40 - 4$

$9 \times 4 = 36$

The break apart and distribute strategy lets me break apart an array that represents a large multiplication fact into arrays that represent smaller facts to find the product. This strategy can be applied in a special way to multiply by 9 by finding 10 units and subtracting 1 unit.

Lesson 20

Use the distributive property to break apart division problems into known facts.

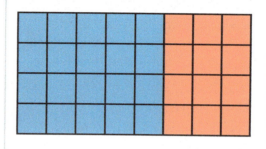

$32 \div 4 = 5 + 3 = 8$

20 12

The break apart and distribute strategy lets me break apart an array that represents a larger division fact that I don't know into smaller arrays that represent division facts that I do know.

Lesson 21

Compose and decompose arrays to create expressions with three factors.

A larger array can be decomposed into equal-size smaller arrays. I represent the total using a multiplication expression with three factors.

Lesson 22

Represent and solve two-step word problems using the properties of multiplication.

a.

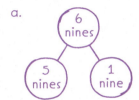

$6 \times 9 = (5 \times 9) + (1 \times 9)$
$6 \times 9 = \quad 45 \quad + \quad 9$
$6 \times 9 = \qquad 54$

Six people spend $54 on lemonade and popcorn.

b.

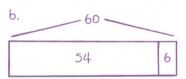

$60 - 54 = 6$

They have $6 left.

To represent and help solve two-step word problems, I make drawings and use strategies that I know.

Lesson 23

Represent and solve two-step word problems using drawings and equations.

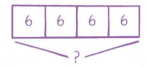

$4 \times 6 = 24$

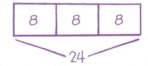

$24 \div 3 = 8$

Each brother gets 8 books.

There are many ways to represent and solve two-step word problems. Seeing the strategies my classmates use helps me to understand their strategies and to learn new strategies that I could try the next time.

19

Use the distributive property to break apart multiplication problems into known facts.

Name _____

✉ **19**

Use the array to complete the equations.

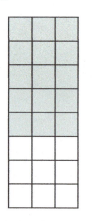

$8 \times 3 = (\underline{\quad 5 \quad} \times 3) + (\underline{\quad 3 \quad} \times 3)$

$8 \times 3 = \underline{\quad 15 \quad} + \underline{\quad 9 \quad}$

$8 \times 3 = \underline{\quad 24 \quad}$

Lesson at a Glance

Students apply the break apart and distribute strategy to breaking apart the rows of an array and representing their work using a series of equations. Students extend the break apart and distribute strategy to multiply by 9 by thinking about $10 - 1$.

Key Questions

- How is using a number bond to represent break apart and distribute similar to using an array? How is it different?

- How do we use tens to help us multiply by nine?

Achievement Descriptors

3.Mod1.AD6 Apply the distributive property to multiply a factor of 2–5 or 10 by another factor. (3.OA.B.5)

3.Mod1.AD8 Multiply and **divide** within 100 fluently with factors 2–5 and 10, recalling from memory all products of two one-digit numbers. (3.OA.C.7)

Agenda

Fluency 10 min

Launch 10 min

Learn 30 min

- Break Apart and Distribute to Multiply
- Break Apart and Distribute 10 Units to Multiply by 9 Units
- Problem Set

Land 10 min

Materials

Teacher

- None

Students

- None

Lesson Preparation

None

Fluency

Counting the Math Way by Threes and Fours

Students construct a number line with their fingers while counting aloud to build fluency with counting by threes and fours and develop a strategy for multiplying.

For each skip-count, show the math way on your fingers while students count, but do not count aloud.

Let's count the math way by threes. Each finger represents 3.

Have students count the math way by threes from 0 to 30 and then back down to 0.

Hands down. Now you count aloud while I show the count on my fingers. Ready?

Lead students to count forward and backward by threes; emphasize counting on from 15.

Repeat the process with fours.

Whiteboard Exchange: Add Within 1,000

Students add within 1,000 to prepare for similar work beginning in module 2.

Display $15 + 13 =$ _____.

Complete the equation.

Give students time to work. When most students are ready, signal for students to show their whiteboards. Provide immediate and specific feedback. If students need to revise, briefly return to validate their corrections.

Show the answer: 28

$$15 + 13 = \underline{\hspace{2cm}}$$

Teacher Note

As students complete each equation, they may notice a pattern. Allow a few students to share how they are adding 3-digit numbers mentally by using place value strategies.

$$15 + 13 = \underline{\;28\;}$$

$$115 + 13 = \underline{\;128\;}$$

$$115 + 113 = \underline{\;228\;}$$

Repeat the process with the following sequence:

$15 + 13 = \underline{\ 28\ }$	$14 + 12 = \underline{\ 26\ }$	$23 + 24 = \underline{\ 47\ }$
$115 + 13 = \underline{\ 128\ }$	$214 + 12 = \underline{\ 226\ }$	$123 + 124 = \underline{\ 247\ }$
$115 + 113 = \underline{\ 228\ }$	$214 + 112 = \underline{\ 326\ }$	$423 + 424 = \underline{\ 847\ }$
	$224 + 112 = \underline{\ 336\ }$	

I Say, You Say: 5 or 2 of a Unit

Students say the value of a number given in unit form to build fluency for using $5 + n$ with the distributive property.

Invite students to participate in I Say, You Say.

When I say a number in unit form, you say its value. Ready?

When I say 5 fives, you say?

25

5 fives

25

5 fives

25

Repeat the process with the following sequence:

5 twos	5 threes	5 fours	2 fives	2 twos	2 threes	2 fours

Launch 10

Students select and explain an efficient strategy to solve an *equal groups with unknown product* word problem.

Direct students to problem 1 in their books. Use the Math Chat routine to engage students in mathematical discourse.

1. Mrs. Smith bakes a tray of mini muffins.

   The tray has 7 rows of 4 mini muffins.

   What is the total number of mini muffins Mrs. Smith bakes?

   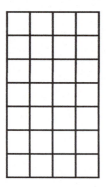

   Mrs. Smith bakes a total of 28 mini muffins.

Give students 2 minutes of silent work time to solve. Have students give a silent signal to indicate they are finished.

Have students discuss their work with a partner. Circulate and listen as they talk. Identify a few students to share their thinking. Purposefully choose work that allows for rich discussion about the strategies students chose and their reasoning. Emphasize thinking that shows the break apart and distribute strategy.

Teacher Note

If a student shares the break apart and distribute strategy by breaking apart the 7 rows into 5 rows and 2 rows, use this to support the transition to Learn by modeling a think aloud such as the following:

"I wonder if we can use the break apart and distribute strategy by breaking apart rows instead of columns. Let's look into this thinking some more."

"I thought about 7 rows of 4 as 7×4. I don't know 7×4, but I can break it apart into two multiplication facts that I do know. I know that 5×4 is 20 and 2×4 is 8, so 7×4 is 28."

Then facilitate a class discussion. Invite students to share their thinking with the whole group, and record their reasoning.

I used the break apart and distribute strategy. When I look at the array, I see columns of 7. Thinking of 4 sevens as 2 sevens and 2 sevens helped me because 2 sevens is 14, and $14 + 14 = 28$.

Ask questions that invite students to make connections, and encourage them to ask questions of their own.

When we use the break apart and distribute strategy, we break apart the columns. We can also use the strategy by breaking apart the rows.

Transition to the next segment by framing the work.

Today, we will break apart the rows of an array to multiply.

Break Apart and Distribute to Multiply

Students solve a multiplication problem by breaking apart the rows in an array and using a number bond to represent the decomposition.

Direct students to problem 2.

2. Miss Diaz's class is going on a field trip.

   The bus has 8 rows of 4 seats.

   How many seats are on the bus?

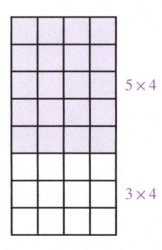

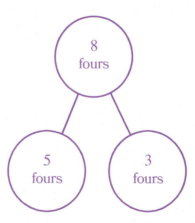

$$8 \times 4 = (5 \times 4) + (3 \times 4)$$
$$= \quad 20 \quad + \quad 12$$
$$= \qquad 32$$

There are 32 seats on the bus.

Guide students through the process of using the break apart and distribute strategy by breaking apart the rows to find out how many seats are on the bus. Consider the following possible sequence:

What multiplication fact represents the entire array?

Write 8×4 below the array.

Let's use the break apart and distribute strategy to break apart the 8 rows of 4. Break the array into 5 rows of 4 and 3 rows of 4 by shading 5 rows of 4 with your pencil.

What multiplication fact represents the shaded part of the array?

Write 5×4 next to the shaded rows in the array.

What multiplication fact represents the unshaded part of the array?

Write 3×4 next to the unshaded rows in the array.

Let's make a number bond to show how we broke apart the 8 fours. What are the shaded and unshaded parts in the array?

5 fours and 3 fours

Draw the number bond as shown in the sample response.

Guide students to complete the equations as shown in the sample response to find 8×4. Conclude by asking the following question.

What is the product of 8 and 4?

Prompt students to write a solution statement. Direct students to rotate their book to show the 8 rows as columns.

Invite students to turn and talk to compare how using the break apart and distribute strategy with rows is similar to using the break apart and distribute strategy with columns.

UDL: Representation

After directing students to break the array into 5 rows of 4 and 3 rows of 4 by shading, consider pausing. Ask students to think about why the 8 rows were broken into 5 rows and 3 rows. Allow students time to generate and share ideas. Review the relationship of 8, 5, and 3. Emphasize that working with fives facts and threes facts is easier than working with eights facts. Pausing here provides time for students to process information and highlights the reasoning involved in using the break apart and distribute strategy.

Promoting the Standards for Mathematical Practice

Students communicate precisely (MP6) when they represent the break apart and distribute strategy using pictorial models and equations.

Ask the following questions to promote MP6:

- How can we write 8×4 using two other multiplication facts?

- How are you using the equations to describe your array (or number bond)?

- How are you using parentheses in your work?

Break Apart and Distribute 10 Units to Multiply by 9 Units

Students multiply by 9 by finding 10 units and subtracting 1 unit.

Direct students to problem 3.

3.

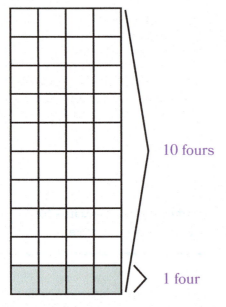

10 fours

1 four

$9 \text{ fours} = 10 \text{ fours} - 1 \text{ four}$

$9 \times 4 = (10 \times 4) - (1 \times 4)$

$9 \times 4 = \quad 40 \quad - \quad 4$

$9 \times 4 = \quad\quad\quad 36$

Guide students through the process of using the break apart and distribute strategy by breaking apart the rows. Consider the following possible sequence:

> **Let's find 9×4. One way to do this quickly in my head is to think about 10×4. We know how to multiply by 10 really well, and 9 is 1 less than 10. Let's use this relationship to help us find 9×4.**

> **Look at the array. How many fours are in the entire array?**

10 fours

Label the array 10 fours.

> **9 fours is 1 four less than 10 fours. So I can think about 9 fours as 10 fours minus 1 four.**

Label the shaded row as 1 four on the array.

Write 9 fours = 10 fours − 1 four.

> **Let's write that as a multiplication equation.**

Write $9 \times 4 = (10 \times 4) - (1 \times 4)$.

Guide students to complete the equations to find 9×4.

Conclude by asking the following question.

> **What is the product of 9 and 4?**

Invite students to think–pair–share about where they see the break apart and distribute strategy being used.

> 10 fours is broken into 9 fours + 1 four. Since I know 10 fours is 40 and 1 four is 4, I can find 9 fours by subtracting: $40 - 4 = 36$.

Teacher Note

Students multiply with 9 as a factor when working with the factors of 2, 3, 4, 5, and 10 throughout module 1. Thinking of 9 as $10 - 1$ supports students when multiplying where one of the factors is 9. This strategy is more efficient than skip-counting by the smaller factor 9 times.

For example, to find 9×4, instead of skip-counting by 4 nine times, I can use tens to help me multiply.

$(10 \times 4) - (1 \times 4)$ or $40 - 4$

This distributive strategy is revisited in module 3, along with explicit instruction of multiples of 9.

Prompt students to work with a partner to use the same strategy to complete problem 4.

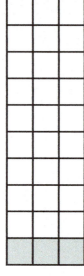

4.

$9 \text{ threes} = 10 \text{ threes} - 1 \text{ three}$

$9 \times 3 = (10 \times 3) - (1 \times 3)$

$9 \times 3 = \quad 30 \quad - \quad 3$

$9 \times 3 = \quad\quad\quad 27$

Invite students to turn and talk about how using 10 can help to multiply by 9.

Problem Set

Differentiate the set by selecting problems for students to finish independently within the timeframe. Problems are organized from simple to complex.

Land 10

Debrief 5 min

Objective: Use the distributive property to break apart multiplication problems into known facts.

Use the following prompts to guide a class discussion on using the break apart and distribute strategy.

How is using the break apart and distribute strategy to break apart the rows of an array similar to using it to break apart the columns of an array? How is it different?

Whether we break apart the rows or the columns of an array, we are breaking the multiplication problem into facts we know so that it is easier to multiply.

It's different because I think about the rows as groups instead of thinking about the columns as groups.

How is using a number bond to represent break apart and distribute similar to using an array? How is it different?

Either way, I break the problem into parts that are easier for me to think about.

I use the number bond when I know how I want to break apart the fact into smaller facts.

I use the array to help me see how to break apart the fact.

How do we use ten to help us multiply by nine?

We can multiply the unit by 10, and then subtract 1 unit to get 9 units.

Exit Ticket 5 min

Provide up to 5 minutes for students to complete the Exit Ticket. It is possible to gather formative data even if some students do not complete every problem.

Sample Solutions

Expect to see varied solution paths. Accept accurate responses, reasonable explanations, and equivalent answers for all student work.

 19

Name _____

Use each array to complete the equations.

1.

$(5 \times 4) =$ ___20___

$(1 \times 4) =$ ___4___

6 fours = 5 fours + 1 four
$6 \times 4 = (5 \times 4) + (1 \times 4)$
$6 \times 4 =$ 20   + ___4___
$6 \times 4 =$ ___24___

2.

$(5 \times 4) =$ ___20___

$(2 \times 4) =$ ___8___

7 fours = 5 fours + 2 fours
$7 \times 4 = (5 \times 4) + (2 \times 4)$
$7 \times 4 =$ 20   + ___8___
$7 \times 4 =$ ___28___

167

Complete each number bond. Then, use it to complete the equations.

3.

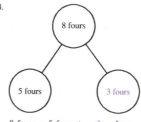

8 fours = 5 fours + ___3___ fours
$8 \times 4 = (5 \times 4) + (\underline{\ 3\ } \times 4)$
$8 \times 4 =$ 20   + ___12___
$8 \times 4 =$ ___32___

4.

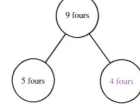

9 fours = 5 fours + ___4___ fours
$9 \times 4 = (5 \times 4) + (\underline{\ 4\ } \times 4)$
$9 \times 4 =$ 20   + ___16___
$9 \times 4 =$ ___36___

---

5. David used the break apart and distribute strategy.

Look at David's work.

What numbers did David multiply? Complete the number bond and the equation.

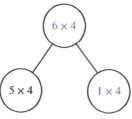

David's work:

$(5 \times 4) + (1 \times 4) = 24$
$20 + 4 = 24$

___6___ × ___4___ = ___24___

Use each array to complete the equations.

6.

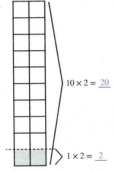

$10 \times 2 = \underline{20}$

$1 \times 2 = \underline{2}$

9 twos = 10 twos − 1 two

$9 \times 2 = (10 \times 2) - (1 \times 2)$

$9 \times 2 = 20 - \underline{2}$

$9 \times 2 = \underline{18}$

7.

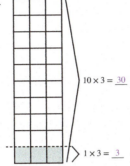

$10 \times 3 = \underline{30}$

$1 \times 3 = \underline{3}$

9 threes = 10 threes − 1 three

$9 \times 3 = (10 \times 3) - (1 \times 3)$

$9 \times 3 = 30 - \underline{3}$

$9 \times 3 = \underline{27}$

Use the Read–Draw–Write process to solve the problem.

8. Luke makes 9 pancakes.

He puts 4 berries on each pancake.

What is the total number of berries that Luke uses?

Complete the number bond. Then, use the break apart and distribute strategy to multiply.

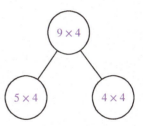

$9 \times 4 = (5 \times 4) + (4 \times 4)$

$9 \times 4 = \quad 20 \quad + \quad 16$

$9 \times 4 = \qquad 36$

Luke uses 36 berries.

20

Use the distributive property to break apart division problems into known facts.

Name _____

✉ **20**

Use the break apart and distribute strategy to find 28 ÷ 4. Explain your thinking.

Sample:

$$28 \div 4 = \underline{5} + \underline{2} = \underline{7}$$

20   8

I broke apart 28 into 20 and 8. I chose 20 because I know that 5 fours is 20. I also know that 2 fours is 8. I added 5 and 2 to get the answer, 7.

Copyright © Great Minds PBC

179

Lesson at a Glance

Students apply the break apart and distribute strategy to find quotients. They represent the strategy by using concrete models, pictorial arrays, and number bonds.

Key Questions

- How are fives facts helpful for breaking apart division problems?

- How is using the break apart and distribute strategy for division similar to using it for multiplication? How is it different?

Achievement Descriptors

3.Mod1.AD7 **Represent** and **explain** division as an unknown factor problem. (3.OA.B.6)

3.Mod1.AD8 **Multiply** and **divide** within 100 fluently with factors 2–5 and 10, recalling from memory all products of two one-digit numbers. (3.OA.C.7)

Agenda

Fluency 10 min

Launch 10 min

Learn 30 min

- Concretely Break Apart the Total to Divide
- Pictorially Break Apart the Total to Divide
- Problem Set

Land 10 min

Materials

Teacher

- Interlocking cubes, 1 cm (28)
- Ruler
- Blue colored pencil
- Red colored pencil

Students

- Interlocking cubes, 1 cm (28 per student pair)
- Ruler (1 per student pair)
- Blue colored pencil
- Red colored pencil

Lesson Preparation

None

Fluency 10

Whiteboard Exchange: Picture Graphs

Students answer a question about a picture graph to maintain measurement concepts from grade 2.

Display the information about the picture graph.

> **Let's read the information together.**

After reading, show the picture graph.

> **Here's Amy's picture graph.**

> **What is the title of the picture graph?**

Give students time to work. When most students are ready, signal for students to show their whiteboards. Provide immediate and specific feedback. If students need to revise, briefly return to validate their corrections.

Repeat the process with the following questions:

> **How many animals are mammals?**

11

> **How many more animals are mammals than fish?**

6

> **How many more animals are mammals and fish than birds and reptiles?**

7

> **How many fewer animals are reptiles than mammals?**

8

Amy makes a picture graph of the types of animals at the zoo.

Birds   Fish   Mammals   Reptiles

Animals at the Zoo

Birds	Fish	Mammals	Reptiles

Key: Each ● stands for 1 animal.

Whiteboard Exchange: Add or Subtract Within 1,000

Students add or subtract within 1,000 to prepare for similar work beginning in module 2.

Display $17 + 12 =$ _____.

Complete the equation.

$$17 + 12 = \underline{\hspace{2cm}}$$

Give students time to work. When most students are ready, signal for students to show their whiteboards. Provide immediate and specific feedback. If students need to revise, briefly return to validate their corrections.

Show the answer: 29

Repeat the process with the following sequence:

$217 + 12 = 229$	$217 + 312 = 529$	$15 + 14 = 29$	$29 - 15 = 14$	$315 + 214 = 529$
$529 - 315 = 214$	$36 + 53 = 89$	$89 - 36 = 53$	$436 + 253 = 689$	$689 - 253 = 436$

Launch ⏱ 10

Materials—S: Interlocking cubes

Students explore efficient strategies to concretely model division as equal sharing.

Provide pairs of students with 28 cubes. Display the problem:

> Liz has 28 cubes to divide equally among 4 students.
> She gives 5 cubes to each student.
> Then she gives 2 cubes to each student.
> David wonders why she used that strategy to divide the cubes.
> What do you think Liz's explanation might be?

Invite students to work with a partner to figure out what Liz's strategy might be.

Circulate and listen as students talk. Look for examples that show reasoning about Liz's work. Purposefully select work that allows for rich discussion about the strategies students choose.

Liz knows 5 fours is 20, so she knew she could give 5 to each of the 4 students. Then she saw that she had 8 cubes left. Because 2 fours is 8, each person got 2 more cubes. She broke the 28 cubes into 20 cubes and 8 cubes.

Then facilitate a class discussion. Invite students to share their thinking with the whole group and record their reasoning.

Transition to the next segment by framing the work.

Today, we will apply Liz's strategy to divide by using facts we already know.

Learn 30

Concretely Break Apart the Total to Divide

Materials—T/S: Interlocking cubes, ruler, red and blue colored pencils

Students divide by breaking apart an array into a fives fact and another fact.

Write $21 \div 3 = \underline{\hspace{1cm}}$.

Interactively model solving the problem while students follow along.

Let's count the math way by threes until we get to 21.

How many threes are in 21?

With your hands, break 7 threes into two parts, like a number bond. What are the two parts?

5 threes and 2 threes

Put your hands together and say 7 threes. Then separate your hands and say 5 threes and 2 threes.

Let's build an array to show 7 threes. Use your cubes to make an array with 7 columns of 3.

Use a ruler to break apart, or decompose, the array into 5 threes and 2 threes.

Model breaking the array with a ruler.

Invite students to turn and talk about how the array and counting the math way show $21 \div 3 = 7$.

> ### Teacher Note
>
> The digital interactive Break-Apart Arrays helps students visualize and work with properties of distributive multiplication.
>
> Consider allowing students to experiment with the tool individually or demonstrating the activity for the whole class.

Let's make a drawing to show how we decomposed the array.

Direct students to problem 1 in their books. Invite students to draw with you as you model tracing the outline of the array on the centimeter grid.

1.

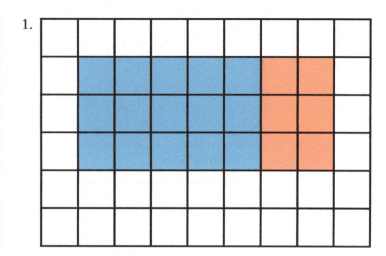

$$21 \div 3 = 7$$

15   6

$$5 + 2 = 7$$

Direct students to color the squares representing the 5 threes blue and the squares representing the 2 threes red. Use a sequence such as the following to interactively model and record the computational thinking.

> **Let's think about how decomposing, or breaking apart, the array can help us find the quotient. How did we decompose the total of 21 into smaller parts? Say it in unit form.**
> 5 threes and 2 threes.

> **What is 5 threes?**

Make a number bond under 21 and write 15 in blue as a part.

> **What is 2 threes?**

Write 6 in red as the other part.

> **We can divide each part of 21 by 3 to help us find the quotient. What is $15 \div 3$?**

Write 5 in blue below the number bond.

What is 6 ÷ 3?

Write 2 in red.

> **Now that we have divided the parts, we can add to find the quotient for 21 ÷ 3.**

> **What is 5 + 2?**

Write + and = 7 in black to complete the addition equation.

> **So, 21 ÷ 3 = 7.**

Invite students to turn and talk about how 5 + 2 = 7 tells us the quotient for 21 ÷ 3.

Pictorially Break Apart the Total to Divide

Materials—S: Red and blue colored pencils

Students break apart an array to find the quotient.

Direct students to problem 2. Interactively model by using the break apart and distribute strategy to divide.

2.

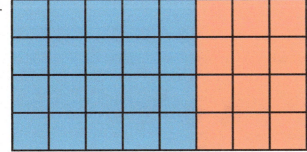

$$32 \div 4 = \textcolor{blue}{5} + \textcolor{orange}{3} = 8$$

$$\textcolor{blue}{20} \quad \textcolor{orange}{12}$$

> **How many total squares are in the array?**

> **Let's use the array and break apart and distribute strategy to find 32 ÷ 4.**

Write 32 ÷ 4 =.

Use a sequence such as the following to model and record the thinking for the selected strategy. Consider simplifying the notation from the previous example.

I know 5 fours is 20. So I'll break 32 into 20 and some other number. 20 and what is 32?

Make a number bond underneath 32, and write 20 in blue as one part and 12 in red as the other part.

How many fours are in 20?

Next to $32 \div 4 =$, write 5 in blue.

How many fours are in 12?

Write $+$ in black and 3 in red.

What is $5 + 3$?

So, what is $32 \div 4$?

Write $= 8$ in black.

To decompose the total, we have been using our fives facts since we know them really well, but those are not the only facts we know well. Is there another way to decompose the total into smaller parts?

Direct students to problem 3. Prompt students to work with a partner to practice another possible way to break apart the total and to show their decomposition on the array provided.

3.

Sample:

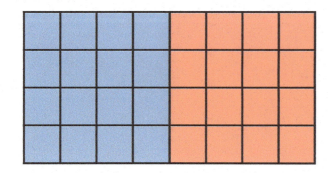

Invite students to share how they divided. If possible, select a student who partitioned the array into 4 fours and 4 fours to share.

Invite students to turn and talk about when the break apart and distribute strategy is helpful for division.

Problem Set

Differentiate the set by selecting problems for students to finish independently within the timeframe. Problems are organized from simple to complex.

Land

Debrief 5 min

Objective: Use the distributive property to break apart division problems into known facts.

Initiate a class discussion by using the prompts below. Encourage students to restate their classmates' responses.

How are fives facts helpful for breaking apart division problems?

Sometimes it is hard to look at a division problem and know what facts it can be broken into. Since I know my fives facts, that is an efficient way for me to start breaking the problem apart.

How is using the break apart and distribute strategy for division similar to using it for multiplication? How is it different?

It is similar because I am using facts I know to help me find the value of facts that I do not know.

It is different because for division I break apart the total, but for multiplication I break apart one of the factors.

Promoting the Standards for Mathematical Practice

Students construct viable arguments and critique the reasoning of others (MP3) when they share their strategies for decomposing the total and breaking apart the array.

Ask the following questions to promote MP3:

- Why does your strategy work? Convince your partner.

- What questions can you ask your partner about why they believe their method was the most efficient?

Exit Ticket 5 min

Provide up to 5 minutes for students to complete the Exit Ticket. It is possible to gather formative data even if some students do not complete every problem.

Sample Solutions

Expect to see varied solution paths. Accept accurate responses, reasonable explanations, and equivalent answers for all student work.

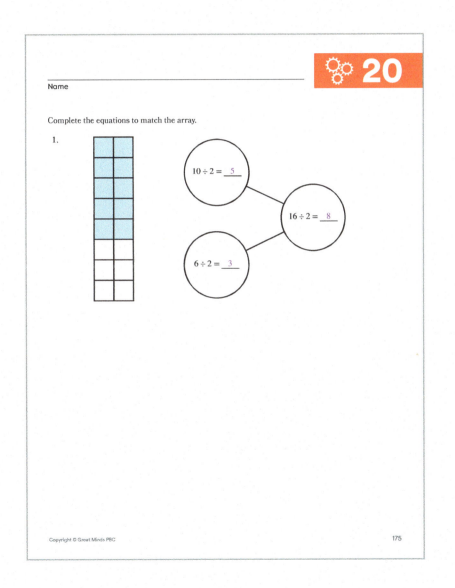

20

Name

Complete the equations to match the array.

1.

$10 \div 2 = \underline{5}$

$16 \div 2 = \underline{8}$

$6 \div 2 = \underline{3}$

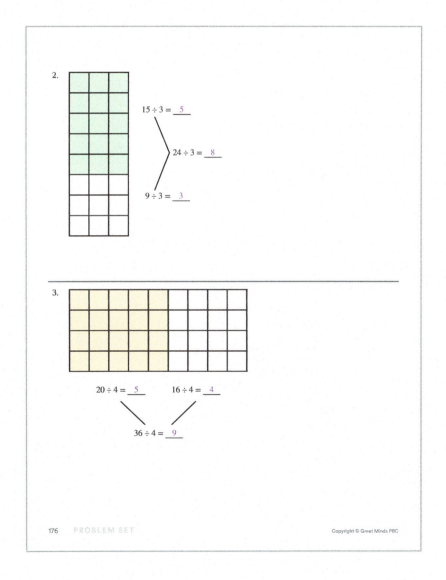

2.

$15 \div 3 = \underline{5}$

$24 \div 3 = \underline{8}$

$9 \div 3 = \underline{3}$

3.

$20 \div 4 = \underline{5}$ $16 \div 4 = \underline{4}$

$36 \div 4 = \underline{9}$

Use the arrays to help you complete the equations.

4.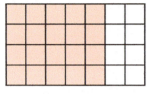

$$28 \div 4 = 5 + \underline{2}$$

20   8

5.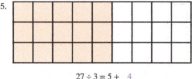

$$27 \div 3 = 5 + \underline{4}$$

15   12

Divide using the break apart and distribute strategy.

6. $$21 \div 3 = 5 + \underline{2} = \underline{7}$$

15   6

7. $$24 \div 4 = \underline{5} + \underline{1} = \underline{6}$$

20   4

8. $$32 \div 4 = \underline{5} + \underline{3} = \underline{8}$$

20   12

9. Use the break apart and distribute strategy to find $36 \div 6$. Explain your thinking.

Sample:

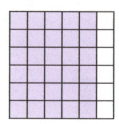

$$36 \div 6 = 5 + 1 = 6$$

30   6

I broke apart 36 into 30 and 6. I chose 30 because I know that 5 sixes is 30. I also know that 1 six is 6. I added 5 and 1 to get the answer, 6.

Compose and decompose arrays to create expressions with three factors.

Name _____

1. Fill in the blanks to match the arrays.

I see __2__ groups of __3__ × __4__ .

__2__ × (__3__ × __4__)

2. Fill in the blanks to match the arrays.

__3__ × (__2__ × __2__) = __12__

Lesson at a Glance

Students establish a foundation for the associative property of multiplication by composing and decomposing arrays into three factors. Students use statements and three-factor multiplication expressions to represent the groups of arrays.

Key Questions

- How is breaking apart an array into three factors different than describing an array with two factors?
- How can thinking about different ways to make equal groups help us multiply more efficiently?

Achievement Descriptors

3.Mod1.AD8 **Multiply** and **divide** within 100 fluently with factors 2–5 and 10, recalling from memory all products of two one-digit numbers. (3.OA.C.7)

Agenda

Fluency 10 min

Launch 5 min

Learn 35 min

- Break Apart an Array into Smaller Arrays
- Break Apart an Array in Many Ways
- Problem Set

Land 10 min

Materials

Teacher

- None

Students

- Interlocking cubes, 1 cm (20 per student)

Lesson Preparation

None

Fluency ⑩

Whiteboard Exchange: Bar Graphs

Students answer a question about a bar graph to maintain measurement concepts from grade 2.

Display the information about the bar graph.

> **Let's read the information together.**

After reading, show the bar graph.

> **Here's Mr. Lopez's bar graph.**

> **What is the title of the bar graph?**

Give students time to work. When most students are ready, signal for students to show their whiteboards. Provide immediate and specific feedback. If students need to revise, briefly return to validate their corrections.

Repeat the process with the following questions:

> **Which animal got the fewest votes?**

Penguin

> **Which animal got the most votes?**

Giraffe

> **How many more students voted for polar bears than penguins?**

3

> **How many students voted for lions or polar bears?**

18

> **How many fewer students voted for lions than giraffes?**

3

Each student in Mr. Lopez's class voted for their favorite zoo animal.

He used the data to make a bar graph.

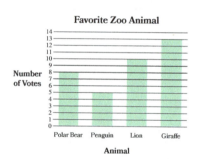

Counting the Math Way by Twos and Fours

Students construct a number line with their fingers while counting aloud to build fluency with counting by twos and fours, and develop a strategy for multiplying.

For each skip-count, show the math way on your fingers while students count, but do not count aloud.

Count the math way by twos from 0 to 20 and then back down to 0 with your partner.

0, 2, 4, 6, 8, 10, 12, 14, 16, 18, 20, (high ten), 20, 18, 16, 14, 12, 10, 8, 6, 4, 2, 0 (double fist bump)

Hands down. Now you count aloud while I show the count on my fingers. Ready?

Lead students to count forward and backward by twos, emphasize counting on from 10.

Now let's count the math way by fours. Each finger represents 4.

Have students count the math way by fours from 0 to 40 and then back down to 0.

Hands down. Now you count aloud while I show the count on my fingers. Ready?

Lead students to count forward and backward by fours; emphasize counting on from 20.

I Say, You Say: 5 or 3 of a Unit

Students say the value of a number given in unit form to build fluency for using $5 + n$ with the distributive property.

Invite students to participate in I Say, You Say.

When I say a number in unit form, you say its value. Ready?

When I say 5 fives, you say?

25

5 fives

25

5 fives

25

Repeat the process with the following sequence:

5 twos	5 threes	5 fours	3 fives	3 twos	3 threes	3 fours

Launch 5

Materials—S: Cubes

Students build equal groups of arrays.

Write the sentence frame: I see _____ groups of _____.

Provide 20 cubes to each student. Invite each student to build an array with 4 rows of 5 cubes on their whiteboards.

Direct students to discuss their arrays with a partner by using the following prompts:

Use the sentence frame to describe your array to your partner.

Tell your partner the multiplication expression that represents your array.

Ask students to share their multiplication expressions. Emphasize thinking that shows both 4×5 and 5×4.

Ivan uses the expression 4×5. Shen uses the expression 5×4. Who is correct?

Invite students to justify how both expressions correctly represent the array.

Both expressions are correct because the commutative property says we can multiply in any order. For this activity, let's think of this array as 4×5.

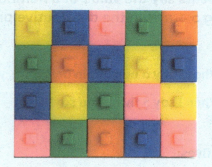

Refer to figure 21.1.

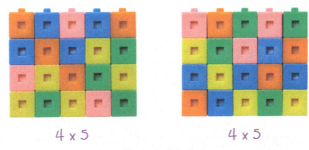

4 x 5   4 x 5

Figure 21.1

Move your array next to your partner's array, but do not let them touch. Under each array, write the multiplication expression that describes your array, starting with the number of rows.

Write the following sentence frame: I see _____ groups of ___ × ___.

Point to the sentence frame and have students think–pair–share about the following questions:

How could we complete this sentence frame to describe our arrays?

I see 2 groups of 4×5.

What other math expressions can we write to describe the arrays?

Have students discuss their work with a partner. Circulate and listen as they talk. Identify a few students to share their thinking. Invite students to share their thinking with the whole group and record their reasoning. Responses may include $4 + 4 + 4 + 4 + 4 + 4 + 4 + 4 + 4 + 4$, $2 \times (4 \times 5)$, and 2×20.

Leave the sentence frame displayed to reference later in the lesson.

Transition to the next segment by framing the work.

Today, we will describe how we see smaller arrays inside larger arrays.

Break Apart an Array into Smaller Arrays

Materials—S: Cubes

Students break an array into three factors.

Invite partners to push their arrays together so they touch to make 4 rows of 10 (see figure 21.2).

> **Now we have 4 rows of 10 cubes. What multiplication equation does your new array represent?**
>
> $4 \times 10 = 40$
>
> **I wonder if there is another way to think about 40 as 2 groups of something. Work with your partner to find another way to show the array in 2 equal groups. Use the sentence frame to describe the groups you made.**

Provide partners time to work. Select partners to share their arrays and sentence frames (see figure 21.3).

Figure 21.2

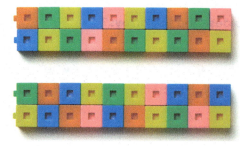

Figure 21.3

Work with your partner to find ways to break apart the arrays to make 4 equal groups. Use the sentence frame, I see _____ groups of ___ × ___, to describe the equal groups.

Provide partners time to work. As partners work, circulate and select examples that show arrays as shown in figures 21.4 and 21.5.

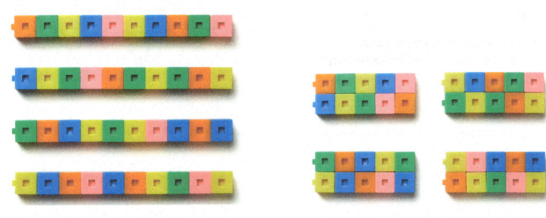

Figure 21.4   Figure 21.5

Invite partners to share each of the cube arrangements and the completed sentence frame. Direct students to the arrangement in figure 21.4.

Let's write a multiplication expression to represent our statement: I see 4 groups of 1×10.

1×10 is repeated 4 times. So, we can write $4 \times (1 \times 10)$. I can use parentheses to group the factors to show how we broke apart the arrays.

Write $4 \times (1 \times 10)$.

Direct students to the arrangement in figure 21.5. Invite students to turn and talk to find the multiplication expression to describe the statement: I see 4 groups of 2×5.

Write $4 \times (2 \times 5)$.

Invite students to turn and talk with their partner about which expression is easier for them to think about and why.

Teacher Note

Encourage students to think flexibly about multiplication expressions that describe an array. This work is preparation for application of the associative property in module 3.

Promoting the Standards for Mathematical Practice

Students look for and express regularity in repeated reasoning (MP8) as they notice patterns that help them break apart the same array into equal groups, e.g., by noticing how the number of rows (or columns) in the original array relates to the number of rows (or columns) in each equal group.

Ask the following questions to promote MP8:

- What patterns do you notice when you compare the number of rows (or columns) in the original array to the number of rows (or columns) in each equal group?

- What patterns do you notice about the equations and products of all the arrays? How could that help you multiply more efficiently?

Break Apart an Array in Many Ways

Materials—S: Cubes

Students break one array into various three-factor expressions.

Invite partners to build an array with 4 rows of 6 cubes.

What multiplication equation does your array represent?

Work with your partner to find various ways to break apart the array into equal groups. Write and complete the sentence frame on your whiteboard to describe each new arrangement.

Provide partners time to work. While partners work, circulate and ask questions such as the following to promote strategic reasoning:

- Can you break the array into a different number of equal groups?
- How could you make a different number of rows? Columns?

> **Teacher Note**
>
> _____
>
> The solutions shown in figure 21.6 are only some of the possible solutions. Accept any student models and explanations that show equal groups.

Some possible solutions are shown in figure 21.6.

2 Groups	3 Groups	4 Groups	6 Groups
I see 2 groups of 4×3. 	I see 3 groups of 4×2. 	I see 4 groups of 2×3. 	I see 6 groups of 4×1. 
I see 2 groups of 2×6. 		I see 4 groups of 1×6. 	

Figure 21.6

Invite partners to share their completed sentence frames and write them where students can study the sentences. Ask questions that invite students to make connections. Consider using the following sequence:

What do you notice?

Three of the sentences have 2, 3, and 4, but they are in different orders.

Two of the sentences have 1, 4, and 6, but they are in different orders.

There are two ways to make 2 groups and 4 groups, but only one way to make 3 groups and 6 groups.

There are many ways to break apart an array with 24 cubes into equal groups of arrays.

Invite students to turn and talk with their partner about which representation of 24 is easiest for them to think about and why.

Direct students to the sentence frames that show 2 groups of 4×3 and 4 groups of 2×3 and ask:

How might it be helpful to think about 2 groups of 4×3 instead of 4 groups of 2×3?

I know my doubles, so I know 2 twelves is 24. But I am still working on learning my fours facts and don't know 4×6 yet.

What product do all our arrays represent?

The product for all the expressions is 24. I can break 24 up into three factors in many different ways.

Problem Set

Differentiate the set by selecting problems for students to finish independently within the timeframe. Problems are organized from simple to complex.

Language Support

Invite students to use a sentence frame to support their discussions. For example:

____ groups of ___ × ___ was the easiest for me to think about because _____.

Differentiation: Challenge

Challenge students to find various ways to break apart an array with 3 rows of 8 cubes into equal groups. Direct students to complete the sentence frame

____ groups of ___ × ___ to describe each arrangement. Consider prompting them to compare the sentences for the 4 rows of 6 cubes and the 3 rows of 8 cubes.

Land  10

Debrief 5 min

Objective: Compose and decompose arrays to create expressions with three factors.

Use the following prompts to guide a class discussion about different ways to break apart arrays.

How is breaking apart an array into three factors different than describing an array with two factors?

Usually we describe how many rows and columns there are in one array. With three factors, we see smaller arrays inside a larger array.

It helps us to see an array in more than one way.

How can thinking about different ways to make equal groups help us multiply more efficiently?

I can make groups that represent facts I know to help me find facts that I don't know.

Exit Ticket 5 min

Provide up to 5 minutes for students to complete the Exit Ticket. It is possible to gather formative data even if some students do not complete every problem.

Sample Solutions

Expect to see varied solution paths. Accept accurate responses, reasonable explanations, and equivalent answers for all student work.

°° 21

Name _____

Fill in the blanks to match the arrays.

1.

I see 2 groups of __3__ × __4__ .

2 × (__3__ × __4__)

2.

I see 3 groups of __2__ × __4__ .

__3__ × (__2__ × __4__)

3.

I see __4__ groups of __3__ × __2__ .

__4__ × (__3__ × __2__)

4. The array is broken apart into 2 equal groups. Fill in the blanks to match the array.

I see 2 groups of __2__ × __5__ .

__2__ × (__2__ × __5__)

PROBLEM SET

5. Draw a line to break apart the array into 2 equal groups.

   Fill in the blanks to match the array.

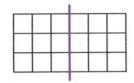

   I see 2 groups of ___3___ × ___3___ .

   ___2___ × (___3___ × ___3___)

6. Fill in the blanks to match the expression.

   Break apart the array by drawing lines to match the expression.

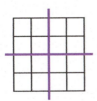

   4 × (2 × 2)

   I see ___4___ groups of ___2___ × ___2___ .

7. The same array is shown four times.

   Show a different way to break apart each array. Then complete the statement.

   Sample:

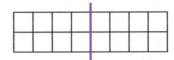

   I see ___2___ groups of ___2___ × ___4___ .

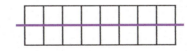

   I see ___2___ groups of ___1___ × ___8___ .

   I see ___4___ groups of ___1___ × ___4___ .

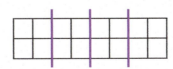

   I see ___4___ groups of ___2___ × ___2___ .

Represent and solve two-step word problems using the properties of multiplication.

Name _____

✉ **22**

Use the Read–Draw–Write process to solve problems (a) and (b).

At lunch, a total of 30 students sit at 5 tables. The same number of students sit at each table.

   a. How many students sit at each table?

      Sample:

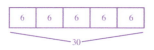

$$30 \div 5 = 6$$

6 students sit at each table.

   b. 4 tables are red, and 1 table is blue. How many students sit at a red table?

      Sample:

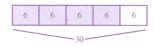

$$4 \times 6 = 24$$

24 students sit at a red table.

Lesson at a Glance

Students select representations and strategies to solve two-step word problems. After working independently to solve the problem, students share their work to compare and connect the strategies. Students make connections between the properties of multiplication that they use to solve problems.

Key Questions

- How can a problem be solved using different representations?
- How can a problem be solved using different equations?

Achievement Descriptors

3.Mod1.AD3 Solve one-step word problems by using multiplication and division within 100, involving factors and divisors 2–5 and 10. (3.OA.A.3)

3.Mod1.AD8 Multiply and **divide** within 100 fluently with factors 2–5 and 10, recalling from memory all products of two one-digit numbers. (3.OA.C.7)

3.Mod1.AD9 Solve two-step word problems. (3.OA.D.8)

Agenda

Fluency 15 min

Launch 5 min

Learn 30 min

- Solve a Two-Step Word Problem
- Share, Compare, and Connect
- Problem Set

Land 10 min

Materials

Teacher

- None

Students

- Count by Twos and Fours Sprint (in the student book)

Lesson Preparation

Consider tearing out the Sprint pages in advance of the lesson.

Fluency

Sprint: Count by Twos and Fours

Materials—S: Count by Twos and Fours Sprint

Students write the unknown number in a sequence to build fluency with counting by twos and fours.

Have students read the instructions and complete the sample problems.

Fill in the blank to complete the sequence.

1.	2, 4, 6, _____	8
2.	36, 32, 28, _____	24

Direct students to Sprint A. Frame the task:

I do not expect you to finish. Do as many problems as you can, your personal best.

Take your mark. Get set. Think!

Time students for 1 minute on Sprint A.

Stop! Underline the last problem you did.

I'm going to read the answers. As I read the answers, call out "Yes!" if you got it correct. If you made a mistake, circle the answer. Ready?

Read the answers to Sprint A quickly and energetically.

Count the number you got correct and write the number at the top of the page. This is your personal goal for Sprint B.

Celebrate students' effort and success.

Provide about 2 minutes to allow students to complete more problems or to analyze and discuss patterns in Sprint A. If students are provided time to complete more problems on Sprint A, reread the answers but do not have them alter their personal goals.

Lead students in one fast-paced and one slow-paced counting activity, each with a stretch or physical movement.

Point to the number you got correct on Sprint A. Remember this is your personal goal for Sprint B.

Direct students to Sprint B.

Take your mark. Get set. Improve!

Time students for 1 minute on Sprint B.

Stop! Underline the last problem you did.

I'm going to read the answers. As I read the answers, call out "Yes!" if you got it correct. If you made a mistake, circle the answer.

Read the answers to Sprint B quickly and energetically.

Count the number you got correct and write the number at the top of the page.

Figure out your improvement score and write the number at the top of the page.

Celebrate students' improvement.

Teacher Note

Consider asking the following questions to discuss the patterns in Sprint A:

- How do problems 1–6 compare to problems 7–12?

- What do you notice about problems 13–22?

Teacher Note

Count forward by fours from 0 to 40 for the fast-paced counting activity.

Count backward by fours from 40 to 0 for the slow-paced counting activity.

Launch ⑤

Students use the properties of multiplication to write equations or statements to represent a picture.

Display the picture of the dice.

Invite students to write equations, expressions, or statements on their whiteboards to describe how the total number of dots are organized. Circulate and observe student work. Select three or four students to share their work. Purposefully choose work that allows for rich discussion about the use of the commutative property, break apart and distribute strategy, or using three factors to multiply.

Examples may include:

- Commutative property: $6 + 6 + 6 + 6 = 24$, $8 \times 3 = 24$, $3 \times 8 = 24$, $4 \times 6 = 24$, and $6 \times 4 = 24$
- Break apart and distribute: $(4 \times 3) + (4 \times 3)$ and $(2 \times 3) + (2 \times 3) + (2 \times 3) + (2 \times 3)$
- Three factors: 4 groups of 2×3, 2 groups of 4×3, and $4 \times (2 \times 3)$

Prompt students to explain where they see their equation, expression, or statement in the picture. If students do not provide some of these examples, consider showing a few and discuss how each relates to the picture. Consider asking the following questions:

- How do the numbers in this equation relate to the picture?
- How is it helpful to think about the total in so many different ways?
- Why can we describe one picture so many ways?

Transition to the next segment by framing the work.

Today, we will solve word problems using familiar strategies.

Learn  30

Solve a Two-Step Word Problem

Students reason about and solve a two-step word problem using self-selected strategies.

Direct students to the problem in their books. Have students work independently to use the Read–Draw–Write process as they solve parts (a) and (b).

> Use the Read–Draw–Write process to solve the problem.
>
> 6 people go to the county fair.
>
> They bring a total of $60 to spend on food.
>
> Each person buys a lemonade and a popcorn.
>
>
>
> Lemonade $4
>
> Popcorn $5
>
> a. How much do the 6 people spend on lemonade and popcorn?
>
> They spend $54 on lemonade and popcorn.
>
> b. How much money do they have left?
>
> They have $6 left.

Circulate and observe student strategies. Select two or three students to share their work in the next segment. Look for work samples that help advance the lesson's objective of using the properties of multiplication to represent and solve two-step word problems. Purposefully choose work that allows for rich discussion about the use of the properties of multiplication students have learned throughout the module.

Teacher Note

A context video for this word problem is available. It may be used to remove language or cultural barriers and provide student engagement. Before directing students to the problem in their books, consider showing the video and facilitating a discussion about what students notice and wonder. This supports students in visualizing the situation before being asked to interpret it mathematically.

UDL: Action & Expression

Consider making available the tools and materials students used in previous lessons, such as interlocking cubes and centimeter grid paper available. This allows students to express learning in flexible ways.

The student work samples demonstrate the use of previously learned multiplication properties.

Break apart the size of the group with a tape diagram.	Break apart the number of groups with a number bond.	Array to show 6×9 as 2 groups of 3×9.
a. $6 \times 9 = \boxed{}$  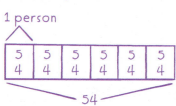 6 fives + 6 fours = 6 nines $(6 \times 5) + (6 \times 4) = 6 \times 9$ $30 + 24 = 54$ The six people spend \$54. b. $60 - 54 = 6$ They have \$6 left.	a. 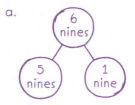 $6 \times 9 = (5 \times 9) + (1 \times 9)$ $6 \times 9 = 45 + 9$ $6 \times 9 = 54$ Six people spend \$54 on lemonade and popcorn. b. 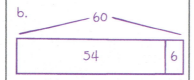 $60 - 54 = 6$ They have \$6 left.	a. 6×9 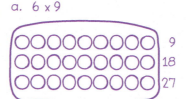 9 / 18 / 27 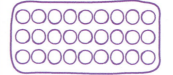 I see 2 groups of 3×9. $3 \times 9 = 27$ $27 + 27 = 54$ Six people spend \$54. b. $60 - 54 = 6$ They have \$6 left.

Promoting the Standards for Mathematical Practice

Students use appropriate tools strategically (MP5) when they choose among tape diagrams, number bonds, equal groups models, equations, and other models, to visualize the parts of each two-step word problem in this lesson.

Ask the following questions to promote MP5:

- What picture or equation could help you see how much each person spends on lemonade and popcorn?
- What picture or equation could help you see how much all 6 people spend on lemonade and popcorn?

Teacher Note

The sample student work shows common responses. Look for similar work from your students and encourage authentic classroom conversations about the key concepts.

If your students do not produce similar work, choose one or two pieces of their work to share and highlight how they show progress toward the goal of this lesson. Then select one sample work from the lesson that would best advance student thinking. Consider presenting the work by saying, "This is how another student solved the problem. What do you think this student did?"

Share, Compare, and Connect

Students share and compare solution strategies for problem 1 and reason about their connections.

Gather the class and invite the students you identified in the previous segment to share their solutions one at a time. Consider purposefully ordering shared student work that shows different solution paths.

As each student shares, ask questions to elicit their thinking and clarify the strategy. Ask the class questions that invite students to make connections between different solution strategies. Encourage students to ask questions of their own.

The sample discussion demonstrates questions that elicit thinking and invite connections.

Break Apart the Size of the Group (Zara's Way)

Zara, how did you represent the problem?

I drew a tape diagram with 6 parts to represent the 6 people.

Explain how Zara's equations relate to the tape diagram.

Her equation shows 6 groups of 5 plus 6 groups of 4. The tape diagram shows how she thought about $5 and $4 for each person.

What does the 6×9 represent?

It represents the total amount spent. There are 6 people and each person spends $9, so $6 \times 9 = 54$.

What does $60 - 54 = 6$ represent?

It represents the amount of money left over. They have $60 and spend $54, so there is $6 left.

Invite students to turn and talk about the similarities and differences between Zara's work and their own work.

Language Support

The Say It Again and Ask for Reasoning sections of the Talking Tool can support students in making connections and asking questions of their own, in this lesson and in lesson 23.

a. $6 \times 9 = \boxed{}$

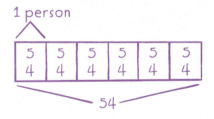

1 person

54

6 fives + 6 fours = 6 nines
$(6 \times 5) + (6 \times 4) = 6 \times 9$
 30 + 24 = 54

The six people spend $54.

b. $60 - 54 = 6$

They have $6 left.

Break Apart the Number of Groups (Adam's Way)

Adam, how did you represent the problem?

I drew a number bond and a tape diagram.

What does 9 represent?

It represents how much money each person spent. Each person buys lemonade for $4 and popcorn for $5, so they spend $9 each.

Why did you break the 6 nines into 5 nines and 1 nine?

I know my fives, so it made it easier to solve.

How did Adam find out they had $6 left?

He drew a tape diagram and he subtracted the amount they spent from the total amount of money they had.

How is Adam's way different from Zara's way?

They used different drawings.

Zara broke apart the 9 and Adam broke apart the 6.

Invite students to turn and talk about the similarities and differences between Adam's work and their own work.

a.

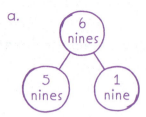

$6 \times 9 = (5 \times 9) + (1 \times 9)$
$6 \times 9 = \quad 45 \quad + \quad 9$
$6 \times 9 = \quad\quad 54$

Six people spend $54 on lemonade and popcorn.

b.

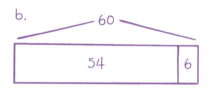

$60 - 54 = 6$

They have $6 left.

See Smaller Arrays Within a Larger Array (Casey's Way)

What did Casey do in her drawing?

She drew an array and circled two groups of 3×9.

What expression can we write to show Casey's two groups of 3×9?

$2 \times (3 \times 9)$

Casey, how did you find out they had $6 left?

I solved the same way that Adam did. I know they had $60 and they spent $54, so I used a number bond to show they had $6 left over.

Invite students to turn and talk about the similarities and differences between Casey's work and their own work.

Write the following expressions: $2 \times (3 \times 9)$, 5 nines + 1 nine, and $(6 \times 5) + (6 \times 4)$.

Have students think–pair–share about how the expressions are related.

They are different ways to write 6×9.

One expression breaks apart the 6 nines into 2 groups of 3 nines. Another expression breaks apart the 6 nines into 5 nines and 1 nine. And another breaks apart the 6 nines into 6 fives and 6 fours.

They're different expressions but they all have the same value.

We've been using different strategies to help us multiply and divide. These strategies help us solve problems efficiently.

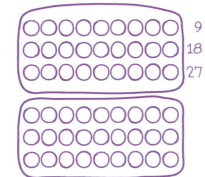

a. 6 x 9

9
18
27

I see 2 groups of 3 x 9.

3 x 9 = 27
27 + 27 = 54

Six people spend $54.

b. 60 – 54 = 6   60

They have
$6 left.   54   6

Problem Set

Differentiate the set by selecting problems for students to finish independently within the timeframe. Problems are organized from simple to complex.

Land

Debrief 5 min

Objective: Represent and solve two-step word problems using the properties of multiplication.

Use the following prompts to guide a discussion about using the properties of multiplication to solve word problems:

How can a problem be solved using different representations?

We can represent problems in ways that make sense to us. We can use tape diagrams, number bonds, arrays, or equal groups.

How can a problem be solved using different equations?

We can break up factors to make different equations that have the same value.

Exit Ticket 5 min

Provide up to 5 minutes for students to complete the Exit Ticket. It is possible to gather formative data even if some students do not complete every problem.

Sample Solutions

Expect to see varied solution paths. Accept accurate responses, reasonable explanations, and equivalent answers for all student work.

A

Number Correct: _____

Fill in the blank to complete the sequence.

1.	0, 2, 4, _____	6
2.	10, 12, 14, _____	16
3.	14, 16, 18, _____	20
4.	6, 4, 2, _____	0
5.	14, 12, 10, _____	8
6.	20, 18, 16, _____	14
7.	0, 4, 8, _____	12
8.	20, 24, 28, _____	32
9.	28, 32, 36, _____	40
10.	12, 8, 4, _____	0
11.	32, 28, 24, _____	20
12.	40, 36, 32, _____	28
13.	4, 6, _____, 10	8
14.	14, _____, 18, 20	16
15.	10, 8, _____, 4	6
16.	6, _____, 2, 0	4
17.	20, _____, 16, 14	18
18.	0, 4, _____, 12	8
19.	28, _____, 36, 40	32
20.	12, 8, _____, 0	4
21.	36, _____, 28, 24	32
22.	40, _____, 32, 28	36

23.	6, 8, _____, 12	10
24.	_____, 14, 16, 18	12
25.	_____, 6, 8, 10	4
26.	_____, 8, 6, 4	10
27.	_____, 10, 8, 6	12
28.	_____, 18, 16, 14	20
29.	20, 24, _____, 32	28
30.	_____, 24, 28, 32	20
31.	_____, 12, 16, 20	8
32.	_____, 28, 24, 20	32
33.	_____, 36, 32, 28	40
34.	_____, 16, 12, 8	20
35.	18, 20, 22, _____	24
36.	28, 26, _____, 22	24
37.	44, 48, 52, _____	56
38.	64, 60, _____, 52	56
39.	36, _____, 40, 42	38
40.	_____, 40, 38, 36	42
41.	72, _____, 80, 84	76
42.	_____, 80, 76, 72	84
43.	_____, 34, 36	32
44.	_____, 84, 80	88

B

Number Correct: _____

Improvement: _____

Fill in the blank to complete the sequence.

1.	2, 4, 6, _____	8
2.	8, 10, 12, _____	14
3.	14, 16, 18, _____	20
4.	8, 6, 4, _____	2
5.	12, 10, 8, _____	6
6.	20, 18, 16, _____	14
7.	4, 8, 12, _____	16
8.	16, 20, 24, _____	28
9.	28, 32, 36, _____	40
10.	16, 12, 8, _____	4
11.	28, 24, 20, _____	16
12.	40, 36, 32, _____	28
13.	2, 4, _____, 8	6
14.	12, _____, 16, 18	14
15.	8, 6, _____, 2	4
16.	8, _____, 4, 2	6
17.	20, _____, 16, 14	18
18.	4, 8, _____, 16	12
19.	24, _____, 32, 36	28
20.	16, 12, _____, 4	8
21.	28, _____, 20, 16	24
22.	40, _____, 32, 28	36

23.	4, 6, _____, 10	8
24.	_____, 12, 14, 16	10
25.	_____, 4, 6, 8	2
26.	_____, 6, 4, 2	8
27.	_____, 8, 6, 4	10
28.	_____, 18, 16, 14	20
29.	12, 16, _____, 24	20
30.	_____, 16, 20, 24	12
31.	_____, 8, 12, 16	4
32.	_____, 20, 16, 12	24
33.	_____, 36, 32, 28	40
34.	_____, 12, 8, 4	16
35.	16, 18, 20, _____	22
36.	26, 24, _____, 20	22
37.	40, 44, 48, _____	52
38.	60, 56, _____, 48	52
39.	34, _____, 38, 40	36
40.	_____, 38, 36, 34	40
41.	68, _____, 76, 80	72
42.	_____, 76, 72, 68	80
43.	_____, 32, 34	30
44.	_____, 80, 76	84

Name _____

Use the Read–Draw–Write process to solve each problem.

1. Carla buys 3 books and 1 card.

   Each book costs $8.

   The card costs $4.

   a. What is the total cost of the books?

   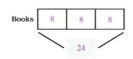

   $3 \times 8 = 24$

   The total cost of the books is $24.

   b. How much does Carla spend altogether?

   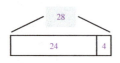

   $24 + 4 = 28$

   Carla spends $28 altogether.

2. 7 students share 28 markers equally.

   a. How many markers does each student get?

   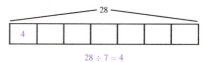

   $28 \div 7 = 4$

   Each student gets 4 markers.

   b. What is the total number of markers shared with 3 of the students?

   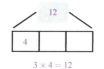

   $3 \times 4 = 12$

   12 markers are shared with 3 of the students.

3. A total of 18 cups are equally packed into 6 boxes.

   a. How many cups are in each box?

   $18 \div 6 = 3$

   There are 3 cups in each box.

   b. All of the cups in 2 boxes are broken. How many cups are not broken?

   $4 \times 3 = 12$

   12 cups are not broken.

　　193

194

4. 25 blue balloons and 15 red balloons are shared equally among 5 children.

    a. **What is the total number of balloons?**

    $$25 + 15 = 40$$

    There are 40 balloons.

    b. **How many balloons does each child get?**

    $$40 \div 5 = 8$$

    Each child gets 8 balloons.

5. Adam packs 27 limes into some bags. Each bag has 3 limes in it.

    a. **How many bags of limes does Adam pack?**

    $$27 \div 3 = 9$$

    Adam packs 9 bags of limes.

    b. **Adam sells 5 of the bags. How many bags are left?**

    $$9 - 5 = 4$$

    4 bags are left.

    c. **How many limes are left?**

    $$4 \times 3 = 12$$

    12 limes are left.

Represent and solve two-step word problems using drawings and equations.

Name _____

✉ 23

Use the Read–Draw–Write process to solve the problem.

Carla buys 5 packs of glow sticks.

Each pack has 8 glow sticks.

Carla uses 12 glow sticks on a project.

How many glow sticks are left?

Sample:

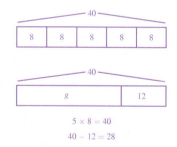

$$5 \times 8 = 40$$
$$40 - 12 = 28$$

There are 28 glow sticks left.

Lesson at a Glance

Students select representations and strategies to solve two-step word problems. After working independently to solve the problem, students share their work to compare and connect solution strategies.

Key Questions

- How do pictorial models help us understand the problems?
- How did you use familiar strategies to solve problems?

Achievement Descriptors

3.Mod1.AD3 Solve one-step word problems by using multiplication and division within 100, involving factors and divisors 2–5 and 10. (3.OA.A.3)

3.Mod1.AD8 Multiply and **divide** within 100 fluently with factors 2–5 and 10, recalling from memory all products of two one-digit numbers. (3.OA.C.7)

3.Mod1.AD9 Solve two-step word problems. (3.OA.D.8)

Agenda

Fluency 10 min

Launch 10 min

Learn 30 min

- Solve a Two-Step Word Problem
- Share, Compare, and Connect
- Problem Set

Land 10 min

Materials

Teacher

- None

Students

- None

Lesson Preparation

None

Fluency 🔟

Whiteboard Exchange: Bar Graphs

Students answer a question about a bar graph to maintain measurement concepts from grade 2.

Display the information about the bar graph.

Let's read the information together.

After reading, show the bar graph.

Here's Jayla's bar graph.

What is the title of the bar graph?

Give students time to work. When most students are ready, signal for students to show their whiteboards. Provide immediate and specific feedback. If students need to revise, briefly return to validate their corrections.

Repeat the process with the following questions:

How many butterflies did Jayla count?

5

How many more bees than grasshoppers were counted?

5

Which bug's count was double the number of grasshoppers?

Spiders

How many bugs did Jayla count at the park?

38

How many fewer butterflies than bees and grasshoppers were counted?

14

Jayla counted bugs at the park.

She used the data to make a bar graph.

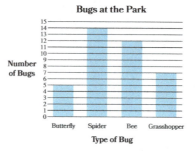

Whiteboard Exchange: Add or Subtract Within $1,000$

Students add or subtract within $1,000$ to prepare for similar work beginning in module 2.

Display $46 + 33 =$ _____.

Complete the equation.

Give students time to work. When most students are ready, signal for students to show their whiteboards. Provide immediate and specific feedback. If students need to revise, briefly return to validate their corrections.

Show the answer: 79

Repeat the process with the following sequence:

$46 + 33 =$ _____

$86 - 53 = 33$	$216 + 313 = 529$	$536 + 343 = 879$	$349 - 123 = 226$	$476 - 135 = 341$

$561 + 434 = 995$	$598 - 256 = 342$	$223 + 24 = 247$	$229 - 15 = 214$

Students use the properties of multiplication to write equations or statements to represent a given picture.

Display the picture of dice. Invite students to write equations, expressions, or statements on their whiteboards to describe how the total number of dots are organized. Circulate and observe student work. Select three or four students to share their work. Purposefully choose work that allows for rich discussion about the use of the commutative property, the break apart and distribute strategy, or using three factors to multiply.

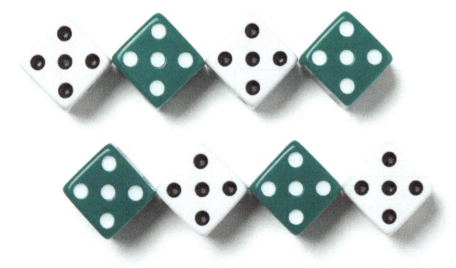

Examples may include:

- Commutative property: $5 + 5 + 5 + 5 + 5 + 5 + 5 + 5 = 40$, $8 \times 5 = 40$, $5 \times 8 = 40$, $4 \times 10 = 40$, and $10 \times 4 = 40$

- Break apart and distribute: $(4 \times 5) + (4 \times 5)$ and $(2 \times 5) + (2 \times 5) + (2 \times 5) + (2 \times 5)$

- Three factors: 4 groups of 2×5, 2 groups of 4×5, $4 \times (2 \times 5)$, and $2 \times (4 \times 5)$

Prompt students to explain where they see their equation, expression, or statement in the picture. If students do not provide some of these examples, consider showing a few and discuss how each relates to the picture. Consider asking the following questions:

- How do the numbers in this equation relate to the picture?
- How is it helpful to think about the total in different ways?
- Why can we describe one picture in so many ways?

Transition to the next segment by framing the work.

Today, we will solve word problems using familiar strategies.

Solve a Two-Step Word Problem

Students reason about and solve a two-step word problem using self-selected strategies.

Direct students to the problem in their books. Have students work independently to use the Read–Draw–Write process to solve the problem. Encourage students to self-select their tools and strategies.

> Use the Read–Draw–Write process to solve the problem.
>
> There are 4 crates with 6 books in each crate.
>
> 3 brothers equally share the books.
>
> How many books does each brother get?
>
> Each brother gets 8 books.

Guide students to reason about the problem by asking questions such as the following:

What information does the problem give us?

What does the question ask?

It asks us how many books each brother gets.

Language Support

An alliteration with the words *books* and *brothers* may present a challenge for students with limited English proficiency. Consider having students act out the problem with props or change the words in the problem to *toys* and *students*.

Promoting the Standards for Mathematical Practice

Students use appropriate tools strategically (MP5) when they choose among tape diagrams, number bonds, equations, and other models, to visualize the parts of each two-step word problem in this lesson.

Ask the following questions to promote MP5:

- What kind of diagram or strategy would be helpful in solving this problem?
- How could a tape diagram or number bond help you find the total number of books?

After a few minutes of work time, invite students to turn and talk about the possibility of solving the problem in one step. Come to a consensus that more than one step is needed. Provide time for students to continue working.

Circulate and observe student work. Select two or three students to share their work in the next segment. Look for work samples that help advance the lesson's objective of using drawings and equations to represent and solve two-step word problems. Purposefully choose work that allows for rich discussion about the use of the properties of multiplication and various solution strategies.

The student work samples shown demonstrate several possible strategies:

Teacher Note

The sample student work shows common responses. Look for similar work from your students and encourage authentic classroom conversations about the key concepts.

If your students do not produce similar work, choose one or two pieces of their work to share and highlight how they show movement toward the goal of this lesson. Then select one sample work from the lesson that would best advance student thinking. Consider presenting the work by saying, "This is how another student solved the problem. What do you think this student did?"

Multiply to find total number of books, then divide to equally share.	Multiply to find total number of books, then use distributive division to equally share.	Divide the 6 books in each crate by the 3 brothers, then multiply the 2 books from each crate by 4 to find how many books each brother gets.
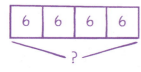 $4 \times 6 = 24$	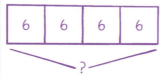 $4 \times 6 = 24$	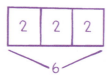 $6 \div 3 = 2$
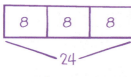 $24 \div 3 = 8$	 $15 \div 3 = 5$ $9 \div 3 = 3$ $5 + 3 = 8$	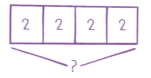 $4 \times 2 = 8$
Each brother gets 8 books.	Each brother gets 8 books.	Each brother gets 8 books.

Share, Compare, and Connect

Students compare solution strategies and reason about their connections.

Gather the class and invite two or three students to share their solutions one at a time. Consider purposefully ordering shared student work that shows different solution paths.

As each student shares, ask questions to elicit their thinking and clarify the strategy they used. Ask the class questions that invite students to make connections between different solution strategies. Encourage students to ask questions of their own.

Multiply then Divide (Robin's Way)

What did Robin do in her drawing?

She drew tape diagrams to show equal groups.

Robin, how did your tape diagrams help you find a solution path?

My first tape diagram shows that I needed to multiply to find the total number of books. My second tape diagram helped me to see that I needed to divide to find how many books each brother gets.

How does Robin's tape show 4×6?

There are 4 parts and each part is 6, so there are 4 sixes, which is the same as 4×6.

How did Robin know to divide 24 by 3?

The problem says 3 brothers equally share the books. So she had to divide the total number of books by 3.

How does the tape diagram show 24 divided by 3?

The entire tape diagram represents the 24 books. The 3 parts represent the 3 brothers. Since Robin knew the total and the number of groups, she divided to find the size of each group. The 8 in each part shows the size of each group.

Invite students to turn and talk about the similarities and differences between Robin's work and their own work.

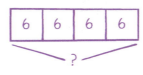

$$4 \times 6 = 24$$

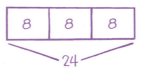

$$24 \div 3 = 8$$

Each brother gets 8 books.

UDL: Action & Expression

After the class compares solution strategies, encourage students to monitor their own progress by prompting them to evaluate the success of their problem-solving approach. For example, provide guiding questions for students to ask themselves, such as:

- How did I do?

- Did I show my thinking?

- Did my strategy work?

- Will I use the same strategy to solve a similar problem next time? Why or why not?

Multiply then Use Distributive Division (Shen's Way)

Shen, tell us what you did to solve the problem.

First, I tried to find the total number of books in the 4 crates. I drew a tape diagram and got a total of 24 books. Then I divided the 24 books equally among the 3 brothers.

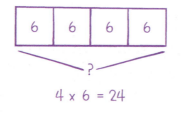

$4 \times 6 = 24$

How did Shen get the quotient of 8?

He broke apart 24 into 15 and 9. Then he divided 15 by 3 and 9 by 3 and added those quotients together.

Invite students to turn and talk to a partner about why Shen is able to add the quotients to get his final answer.

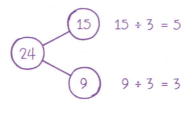

$15 \div 3 = 5$

$9 \div 3 = 3$

$5 + 3 = 8$

Each brother gets 8 books.

Divide then Multiply (Pablo's Way)

How did Pablo think about the problem?

He found out how many books each brother gets from each crate, and then he multiplied his answer by the number of crates.

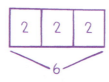

$6 \div 3 = 2$

How does his tape diagram show that?

He drew a tape diagram to represent the 6 books in 1 crate. Then he partitioned it into 3 parts, 1 part to represent each brother. He wrote $6 \div 3 = 2$ to find that each brother gets 2 books from each crate.

Tell us the next step you took to solve, Pablo.

Since each brother gets 2 books from each crate and there are 4 crates, I can multiply. I found that each brother gets 8 books.

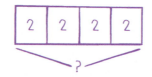

$4 \times 2 = 8$

Each brother gets 8 books.

Invite students to turn and talk to a partner about how Pablo solved the first part and how it is different from what Robin and Shen did.

Why didn't Pablo need to find the total number of books first?

He found the number of books each brother gets from each crate instead of how many each brother gets from all the crates.

Invite students to turn and talk about what is valuable about sharing different ways to solve the same problem.

Problem Set

Differentiate the set by selecting problems for students to finish independently within the timeframe. Problems are organized from simple to complex.

Land 10

Debrief 5 min

Objective: Represent and solve two-step word problems using drawing and equations.

Facilitate a discussion using the following prompts.

How do pictorial models help us understand the problems?

Drawing models helps make sense of the problem and helps us see how many steps there are to solve.

How did you use familiar strategies to solve problems?

I broke apart the numbers to help make easier problems when I was dividing.

Which tool or strategy did you see today that you'd like to try? Why?

Shen used break apart and distribute to divide. I'd like to try that because it breaks a division problem into two simpler problems.

Display *Flower Vendor*, 1949, by Diego Rivera.

Let's look at the painting called *Flower Vendor* by Diego Rivera again.

Use the following questions to help students engage with the art.

- What do you notice today that you didn't see the first time we studied *Flower Vendor*?

- What is something new you wonder about?

- I wonder how heavy the flowers are. Does the woman seem to be struggling to carry the flowers? How do you think she is able to hold so many at once?

Help students relate the art to the concepts developed in module 1. Consider guiding the discussion using the following:

- I wonder how many bundles of flowers the woman has on her back. How many do you think it is?

- How could this painting relate to multiplication? Division?

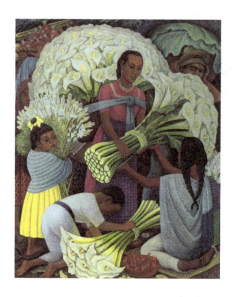

Teacher Note

In the first topic of module 2, students measure the weight of objects in grams and kilograms. *Flower Vendor* provides an authentic opportunity for students to wonder about weight and to consider why it matters.

Exit Ticket 5 min

Provide up to 5 minutes for students to complete the Exit Ticket. It is possible to gather formative data even if some students do not complete every problem.

Sample Solutions

Expect to see varied solution paths. Accept accurate responses, reasonable explanations, and equivalent answers for all student work.

Name _____

Use the Read–Draw–Write process to solve each problem.

1. Mr. Lopez buys 4 packs of 7 markers.

   After he gives 1 marker to each student in his class, he has 5 markers left.

   How many total markers does Mr. Lopez give to his students?

   Step 1: Find the total number of markers that Mr. Lopez buys.

   $$4 \times 7 = 28$$

   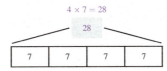

   Step 2: Find the total number of markers Mr. Lopez gives to his students.

   $$28 - 5 = 23$$

   Mr. Lopez gives a total of __23__ markers to his students.

2. Amy has 21 meters of ribbon.

   She cuts the ribbon so that each piece measures 3 meters in length.

   a. How many pieces of ribbon does Amy have?

   $$21 \div 3 = 7$$

   Amy has 7 pieces of ribbon.

   b. If Amy needs a total of 12 pieces, how many more pieces does she need?

   $$12 - 7 = 5$$

   Amy needs 5 more pieces of ribbon.

3. Travis earns money doing chores.

   He earns $6 each week for 4 weeks.

   He earns $4 the fifth week.

   How much does Travis earn altogether?

   $$4 \times 6 = 24$$
   $$24 + 4 = 28$$

   Travis earns $28 altogether.

4. Ivan has a bag of 18 fruit snacks.

   There is an equal number of peach, cherry, and grape fruit snacks.

   Ivan eats all the grape fruit snacks.

   How many fruit snacks does Ivan have left?

   $$18 \div 3 = 6$$
   $$18 - 6 = 12$$

   Ivan has 12 fruit snacks left.

Standards

Module Content Standards

Understand place value.

2.NBT.A.2 Count within 1000; skip-count by 5s, 10s, and 100s.

Represent and solve problems involving multiplication and division.

3.OA.A.1 Interpret products of whole numbers, e.g., interpret 5×7 as the total number of objects in 5 groups of 7 objects each. *For example, describe a context in which a total number of objects can be expressed as 5×7.*

3.OA.A.2 Interpret whole-number quotients of whole numbers, e.g., interpret $56 \div 8$ as the number of objects in each share when 56 objects are partitioned equally into 8 shares, or as a number of shares when 56 objects are partitioned into equal shares of 8 objects each. *For example, describe a context in which a number of shares or a number of groups can be expressed as $56 \div 8$.*

3.OA.A.3 Use multiplication and division within 100 to solve word problems in situations involving equal groups, arrays, and measurement quantities, e.g., by using drawings and equations with a symbol for the unknown number to represent the problem.[1]

3.OA.A.4 Determine the unknown whole number in a multiplication or division equation relating three whole numbers. *For example, determine the unknown number that makes the equation true in each of the equations $8 \times ? = 48$, $5 = \_\_ \div 3$, $6 \times 6 = ?$.*

Understand properties of multiplication and the relationship between multiplication and division.

3.OA.B.5 Apply properties of operations as strategies to multiply and divide.[2] *Examples: If $6 \times 4 = 24$ is known, then $4 \times 6 = 24$ is also known. (Commutative property of multiplication.) $3 \times 5 \times 2$ can be found by $3 \times 5 = 15$, then $15 \times 2 = 30$, or by $5 \times 2 = 10$, then $3 \times 10 = 30$. (Associative property of multiplication.) Knowing that $8 \times 5 = 40$ and $8 \times 2 = 16$, one can find 8×7 as $8 \times (5 + 2) = (8 \times 5) + (8 \times 2) = 40 + 16 = 56$. (Distributive property.)*

---

1   See Glossary, Table 2.

2   Students need not use formal terms for these properties.

3.OA.B.6   Understand division as an unknown-factor problem. *For example, find* $32 \div 8$ *by finding the number that makes* 32 *when multiplied by* 8.

Multiply and divide within 100.

3.OA.C.7   Fluently multiply and divide within 100, using strategies such as the relationship between multiplication and division (e.g., knowing that $8 \times 5 = 40$, one knows $40 \div 5 = 8$) or properties of operations. By the end of grade 3, know from memory all products of two one-digit numbers.

Solve problems involving the four operations, and identify and explain patterns in arithmetic.

3.OA.D.8   Solve two-step word problems using the four operations. Represent these problems using equations with a letter standing for the unknown quantity. Assess the reasonableness of answers using mental computation and estimation strategies including rounding.[3]

Standards for Mathematical Practice

MP1   Make sense of problems and persevere in solving them.

MP2   Reason abstractly and quantitatively.

MP3   Construct viable arguments and critique the reasoning of others.

MP4   Model with mathematics.

MP5   Use appropriate tools strategically.

MP6   Attend to precision.

MP7   Look for and make use of structure.

MP8   Look for and express regularity in repeated reasoning.

---

3   This standard is limited to problems posed with whole numbers and having whole number answers; students should know how to perform operations in the conventional order when there are no parentheses to specify a particular order (Order of Operations).

3.Mod1.AD1 **Represent** a multiplication situation with a model and **convert** between several representations of multiplication.

Note: This excludes the creation of a multiplication situation from an expression, an equation, or a model, which is reserved for module 3.

RELATED CCSSM

3.OA.A.1 Interpret products of whole numbers, e.g., interpret 5×7 as the total number of objects in 5 groups of 7 objects each. *For example, describe a context in which a total number of objects can be expressed as 5×7.*

Partially Proficient	Proficient	Highly Proficient
Represent a multiplication situation **with equal groups or an array**. *Liz arranges her rocks into 3 rows of 10. Draw a model to represent Liz's rocks.*	**Convert** between several representations of multiplication **(e.g., from a situation to a model, an expression, and/or an equation)**. *Complete the equation to describe the picture.* _____ × _____ = _____ *Ivan has 4 bags of acorns. There are 6 acorns in each bag. Write a multiplication expression that could be used to find the total number of acorns Ivan has.*	

3.Mod1.AD2 **Represent** a division situation with a model and **convert** between several representations of division.

Note: This excludes the creation of a division situation from an expression, an equation, or a model, which is reserved for module 3.

RELATED CCSSM

3.OA.A.2 Interpret whole-number quotients of whole numbers, e.g., interpret $56 \div 8$ as the number of objects in each share when 56 objects are partitioned equally into 8 shares, or as a number of shares when 56 objects are partitioned into equal shares of 8 objects each. *For example, describe a context in which a number of shares or a number of groups can be expressed as $56 \div 8$.*

Partially Proficient	**Proficient**	**Highly Proficient**
Represent a division situation **with equal groups or an array**. *Circle groups of pencils to show 20 pencils divided equally into 5 groups.* 	**Convert** between several representations of division **(e.g., from a situation to a model, an expression, or an equation)**. *Complete the equation to describe the picture.* _____ ÷ _____ = _____ *Luke arranges 15 crayons into 3 equal rows. Write a division expression that could be used to find the number of crayons in each row.*	

3.Mod1.AD3 **Solve** one-step word problems by using multiplication and division within 100, involving factors and divisors 2–5 and 10.

Note: Only one factor needs to be 2–5 or 10.

RELATED CCSSM

3.OA.A.3 Use multiplication and division within 100 to solve word problems in situations involving equal groups, arrays, and measurement quantities, e.g., by using drawings and equations with a symbol for the unknown number to represent the problem.[1]

[1] See [CCSSM] Glossary, Table 2.

Partially Proficient	Proficient	Highly Proficient
Solve one-step **equal groups of objects** word problems by using multiplication and division within 100 involving factors and divisors 2–5 and 10. *Ray has 4 boxes of pens. Each box has 8 pens. How many pens does Ray have in all?* *Robin divides 12 cookies equally into 4 bags. How many cookies are in each bag?*	**Solve** one-step **arrays of objects** word problems by using multiplication and division within 100 involving factors and divisors 2–5 and 10. *Shen arranges his shells into 3 rows with 5 shells in each row. How many shells does Shen have?* *Amy arranges 40 rocks into rows of 5. How many rows does she make?*	

3.Mod1.AD4 **Determine** the unknown number in a multiplication or division equation involving factors and divisors 2–5 and 10.

Note: Only one factor needs to be 2–5 or 10.

RELATED CCSSM

3.OA.A.4 Determine the unknown whole number in a multiplication or division equation relating three whole numbers. *For example, determine the unknown number that makes the equation true in each of the equations* $8 \times ? = 48$, $5 = \underline{\hspace{1cm}} \div 3$, $6 \times 6 = ?$

Partially Proficient	Proficient	Highly Proficient
Determine the unknown number in a multiplication or division equation involving factors and divisors 2–5 and 10 **when the unknown is the product or quotient.** *Multiply or divide.* 6×5 $8 \div 2$	**Determine** the unknown number in a multiplication or division equation involving factors and divisors 2–5 and 10 **when the unknown is in any position.** *Fill in the blanks to make true equations.* $\underline{\hspace{1cm}} \times 3 = 12$ $15 \div \underline{\hspace{1cm}} = 5$	

3.Mod1.AD5 **Apply** the commutative property of multiplication to multiply a factor of 2–5 or 10 by another factor.

RELATED CCSSM

3.OA.B.5 Apply properties of operations as strategies to multiply and divide.² *Examples: If $6 \times 4 = 24$ is known, then $4 \times 6 = 24$ is also known. (Commutative property of multiplication.) $3 \times 5 \times 2$ can be found by $3 \times 5 = 15$, then $15 \times 2 = 30$, or by $5 \times 2 = 10$, then $3 \times 10 = 30$. (Associative property of multiplication.) Knowing that $8 \times 5 = 40$ and $8 \times 2 = 16$, one can find 8×7 as $8 \times (5 + 2) = (8 \times 5) + (8 \times 2) = 40 + 16 = 56$. (Distributive property.)*

² Students need not use formal terms for these properties.

Partially Proficient	Proficient	Highly Proficient
Apply the commutative property of multiplication **to generate equivalent expressions**. *Fill in the blanks to make true equations.* $2 \times 7 = \underline{} \times 2$ $5 \times \underline{} = 4 \times 5$ $\underline{} \times 3 = 3 \times 6$	**Explain** the commutative property of multiplication. *Explain why the array represents 2×8 and 8×2.* ○○○○○○○○ ○○○○○○○○	

3.Mod1.AD6 Apply the distributive property to multiply a factor of 2–5 or 10 by another factor.

RELATED CCSSM

3.OA.B.5 Apply properties of operations as strategies to multiply and divide.[2] *Examples: If* $6 \times 4 = 24$ *is known, then* $4 \times 6 = 24$ *is also known. (Commutative property of multiplication.)* $3 \times 5 \times 2$ *can be found by* $3 \times 5 = 15$*, then* $15 \times 2 = 30$*, or by* $5 \times 2 = 10$*, then* $3 \times 10 = 30$*. (Associative property of multiplication.) Knowing that* $8 \times 5 = 40$ *and* $8 \times 2 = 16$*, one can find* 8×7 *as* $8 \times (5 + 2) = (8 \times 5) + (8 \times 2) = 40 + 16 = 56$*. (Distributive property.)*

[2] Students need not use formal terms for these properties.

Partially Proficient	Proficient	Highly Proficient
Apply the distributive property **to generate equivalent expressions**. *Is each expression equal to* 6×5*?* *Circle Yes or No.* $(4 \times 5) + (2 \times 5)$ Yes No $(2 \times 5) + (3 \times 5)$ Yes No $(4 \times 5) \times (2 \times 5)$ Yes No $(5 \times 5) + (1 \times 5)$ Yes No	**Apply** the distributive property to multiply a factor of 2–5 or 10 by another factor. *Break apart the* 8 *to find* 8×4*.* $8 \times 4 = (\underline{} \times 4) + (\underline{} \times 4)$ $= \underline{} + \underline{}$ $= \underline{}$	**Explain** the distributive property for multiplication. *Carla says she can find* 16×5 *by using the expression* $(10 \times 5) + (6 \times 5)$*. Is she correct? Explain.*

3.Mod1.AD7 **Represent** and **explain** division as an unknown factor problem.

RELATED CCSSM

3.OA.B.6 Understand division as an unknown-factor problem. *For example, find* $32 \div 8$ *by finding the number that makes* 32 *when multiplied by* 8.

Partially Proficient	Proficient	Highly Proficient
Recognize related multiplication and division equations. *Which equation can be used to find* $30 \div 5$*?* A. $5 \times \underline{\quad} = 30$ B. $\underline{\quad} \div 5 = 30$ C. $30 \times \underline{\quad} = 5$ D. $30 \times 5 = \underline{\quad}$	**Represent** division as an unknown factor problem by **using equations**. *Pablo has* 18 *fish. He divides them equally into* 3 *bowls. How many fish are in each bowl?* *Write a multiplication equation and a division equation to describe the problem. Use a blank to represent the unknown.*	**Explain** division as an unknown factor problem. *Eva uses the equation* $5 \times \underline{\quad} = 40$ *to find* $40 \div 5$. *Is her thinking correct? Explain.*

3.Mod1.AD8 **Multiply** and **divide** within 100 fluently with factors 2–5 and 10, recalling from memory all products of two one-digit numbers.

Note: Only one factor needs to be 2–5 or 10.

RELATED CCSSM

3.OA.C.7 Fluently multiply and divide within 100, using strategies such as the relationship between multiplication and division (e.g., knowing that $8 \times 5 = 40$, one knows $40 \div 5 = 8$) or properties of operations. By the end of grade 3, know from memory all products of two one-digit numbers.

Partially Proficient	Proficient	Highly Proficient
Multiply and **divide** within 100 fluently with factors 2–5 or 10, **by using the relationship between multiplication and division**.	**Multiply** and **divide** within 100 fluently with factors 2–5 or 10 by recalling from memory all products of two one-digit numbers.	**Multiply** and **divide** within 100 fluently with factors 2–5 or 10 by recalling from memory all products of two one-digit numbers **and a related division fact**.
Use the given equation to fill in the blanks.	*Multiply or divide.*	*Multiply.*
Part A	$20 \div 4$	$2 \times 9 = $ ____
$8 \times 5 = 40$	7×3	*Write a related division equation.*
$40 \div 5 = $ ____		
Part B		
$30 \div 3 = 10$		
$3 \times 10 = $ ____		
$10 \times 3 = $ ____		

3.Mod1.AD9 **Solve** two-step word problems.

Note: For module 1, in multiplication or division problem types, at least one factor or the divisor must be 2–5 or 10.

RELATED CCSSM

3.OA.D.8 Solve two-step word problems using the four operations. Represent these problems using equations with a letter standing for the unknown quantity. Assess the reasonableness of answers using mental computation and estimation strategies including rounding.[3]

[3] This standard is limited to problems posed with whole numbers and having whole-number answers; students should know how to perform operations in the conventional order when there are no parentheses to specify a particular order (Order of Operations).

Partially Proficient	Proficient	Highly Proficient
Solve two-step word problems **involving grades K and 1 addition or subtraction problem types**[1] **and/or equal groups of objects multiplication or division problem types.** *Oka has 14 balloons. She gives 2 balloons to each of her 3 friends. How many balloons does Oka have left?*	**Solve** two-step word problems **involving grade 2 addition or subtraction problem types**[2] **and/or arrays of objects multiplication or division problem types.** *Gabe and Deepa are planting rows of sunflowers. Gabe plants 6 fewer sunflowers than Deepa. He plants 6 rows of 4 sunflowers. How many sunflowers does Deepa plant?*	

[1]   Common Core Standards Writing Team, Progressions for the Common Core, 2011–2015.

[2]   Common Core Standards Writing Team, Progressions for the Common Core, 2011–2015.

Terminology

The following terms are critical to the work of grade 3 module 1. This resource groups terms into categories called New, Familiar, and Academic Verbs. The lessons in this module incorporate terminology with the expectation that students work toward applying it during discussions and in writing.

Items in the New category are discipline-specific words that are introduced to students in this module. These items include the definition, description, or illustration as it is presented to students. At times, this resource also includes italicized language for teachers that expands on the wording used with students.

Items in the Familiar category are discipline-specific words introduced in prior modules or in previous grade levels.

Items in the Academic Verbs category are high-utility terms that are used across disciplines. These terms come from a list of academic verbs that the curriculum strategically introduces at this grade level.

Visit the digital glossary to see complete definitions, illustrations, or descriptions for New and Familiar terms across grade levels. The digital glossary also includes the Academic Verbs list for each grade.

New

commutative property of multiplication
Changing the order of the factors in a multiplication expression does not change its product. For example, if $7 \times 3 = 21$ is known then $3 \times 7 = 21$ is also known because $7 \times 3 = 3 \times 7$. (Lesson 10)

division, divide, divided by, $\div$
To partition a total into groups of a specific size or into a specific number of equal groups. For example, 15 divided by 3 (written $15 \div 3$) can mean either of the following: 15 put into groups of 3 or 15 put into 3 equal groups. (Lessons 7 and 8)

factor
The numbers used in a multiplication expression. For example, in 3×4, the numbers 3 and 4 are the factors. (Lesson 4)

multiplication, multiply, $\times$
The multiplication expression 3×5 means 3 groups of 5. To multiply two whole numbers is to find the total, or product, for a multiplication expression. (Lesson 2)

parentheses, $(\,)$
Symbols used for grouping and ordering parts of an expression. For example, in $3 \times (5 + 1)$, the $5 + 1$ is grouped between the parentheses. The parentheses make it clear that the 3 is being multiplied by $5 + 1$, not just by 5. (Lesson 12)

In grade 3, students gain the general understanding that operations inside parentheses are done before operations outside parentheses. Fluency with parentheses begins in grade 5 and is made formal in grade 6.

product
The total when one number is multiplied by another. For example, in $3 \times 4 = 12$, the number 12 is the product. (Lesson 3)

quotient
The number resulting from the division of two numbers. For example, in $28 \div 4 = 7$, the number 7 is the quotient. (Lesson 15)

rotate

To turn, used in reference to turning arrays 90 degrees. When an array is rotated, the rows become columns and the columns become rows. (Lesson 10)

The degree measure of an angle is introduced to students in grade 4.

size of group

The number in the group. For example, the multiplication equation $3 \times 4 = 12$ represents 3 groups of 4; so, the size of each group is 4. (Lesson 12)

Familiar

array

column

equal groups

equal shares

equal sharing

equation

estimate (noun)

estimate (verb)

expression

number in (or size of) each group

number of groups

repeated addition

row

skip-count

unit

unit form

unknown

Academic Verbs

Module 1 does not introduce any academic verbs from the grade 3 list.

Math Past

Who first used the symbol × for multiplication?
Were there other multiplication symbols?
Do we still use any of them today?

Tell your class that you will write 3 times 5 on the whiteboard. Your students probably expect you to write the following expression:

$$3 \times 5$$

That is, after all, the way modern elementary students learn to write multiplication. In fact, the × symbol to represent multiplication has been in use for over 400 years. It was first used in print in the early 1600s by English mathematician William Oughtred. But along the way, scholars tried, and eventually discarded, several other notations for multiplication.

Instead of writing what students expect, build wonder by writing this:

$$3 \text{ M } 5$$

Tell students that mathematicians in mid-1500s Germany represented multiplication by using an *M* instead of the × symbol. Ask your students to guess why mathematicians picked the letter *M* to represent multiplication. Some students may observe that *M* is the first letter in the word *multiply*. In fact, the *M* stands for the German word *multiplizieren*, meaning "to multiply."

Again, build wonder by writing a different notation:

$$\boxed{}\, 3,\, 5$$

This should start some interesting student discussion. What could the rectangle mean? If it is supposed to mean "multiply," then why isn't it between the 3 and the 5? And why is there a comma?

This way of writing multiplication was introduced in France in the mid-1600s. The rectangular symbol is a shortcut way of saying "find the area of an array containing 3 rows of 5." The comma is actually the multiplication symbol.

Once more, build wonder by writing yet another multiplication notation. Students might be quite amused by this one.

$$3 \cap 5$$

What do students think of the symbol between 3 and 5? Some may see it as a letter *C* drawn sideways or even a pair of headphones.

This way of writing multiplication was invented by mathematician Gottfried Leibniz in 1666. It was Leibniz's first attempt to invent a symbol for multiplication. We do not know why he chose this particular symbol, but it was indeed just a letter *C* drawn sideways. Over the years, Leibniz experimented with other symbols for multiplication and finally discarded the sideways *C* in favor of using a simple dot.

$$3 \cdot 5$$

We still use the dot today, but students have not yet seen it at this stage of learning. In algebra, the dot helps students avoid confusion between the variable x and the multiplication symbol ×. Leibniz even predicted this problem. He wrote,

> I do not like × as a symbol for multiplication, as it is easily confounded with x. ...

Challenge your students to create a multiplication symbol of their own!

Materials

The following materials are needed to implement this module. The suggested quantities are based on a class of 24 students and one teacher.

1	100-bead demonstration rekenrek	60	Paper plates, small
4	Color tiles, plastic, set of 400	25	Pencils
25	Colored pencils, set of 8	25	Personal whiteboards
1	Computer with internet access	25	Personal whiteboard erasers
240	Crackers, round	1	Projection device
25	Dry-erase markers	25	Scissors
14	Envelopes	144	Sticky notes
1,300	Interlocking cubes, 1 cm	1	*Teach* book
24	*Learn* books	25	Wood rulers, inch and metric

Visit http://eurmath.link/EurekaMaterials to learn more.

Please see lesson 1 for a list of organizational tools (cups, rubber bands, graph paper, etc.) suggested for the counting collection.

Works Cited

Cajori, Florian. *A History of Mathematical Notations*, Vol. 2. La Salle, Ill.: The Open Court Publishing Company, 1929.

CAST. *Universal Design for Learning Guidelines version 2.2.* Retrieved from http://udlguidelines.cast.org, 2018.

Common Core Standards Writing Team. *Progressions for the Common Core State Standards in Mathematics*. Tucson, AZ: Institute for Mathematics and Education, University of Arizona, 2011–2015. http://math.arizona.edu/~ime/progressions.

National Governors Association Center for Best Practices, Council of Chief State School Officers (NGA Center and CCSSO). *Common Core State Standards for Mathematics*. Washington, DC: National Governors Association Center for Best Practices, Council of Chief State School Officers, 2010.

Zwiers, Jeffrey, Jack Dieckmann, Sara Rutherford-Quach, Vinci Daro, Renee Skarin, Steven Weiss, and James Malamut. *Principles for the design of mathematics curricula: Promoting language and content development*. Retrieved from Stanford University, UL/SCALE website: http://ell.stanford.edu/content/mathematics-resources-additional-resources, 2017.

Credits

Great Minds® has made every effort to obtain permission for the reprinting of all copyrighted material. If any owner of copyrighted material is not acknowledged herein, please contact Great Minds for proper acknowledgment in all future editions and reprints of this module.

Common Core State Standards for Mathematics © Copyright 2010 National Governors Association Center for Best Practices and Council of Chief State School Officers. All rights reserved.

For a complete list of credits, visit http://eurmath.link /media-credits.

Cover, Paul Klee, (1879–1940), *Farbtafel "qu 1"* (Colour table "Qu 1"), 1930, 71. pastel on coloured paste on paper on cardboard, 37.3 x 46.8 cm. Kunstmuseum Basel, Kupferstichkabinett, Schenkung der Klee-Gesellschaft, Bern. © 2020 Artists Rights Society (ARS), New York.; pages 29, 370, Diego Rivera (1886–1957), *Vendedora de flores* (*Flower Vendor*), 1949. 20th century. Madrid, Reina Sofia museum. © 2020 Banco de México Diego Rivera Frida Kahlo Museums Trust, Mexico, D.F/Artists Rights Society (ARS), New York. Photo credit: Album/ Art Resource, NY; page 208, (left) Mike Flippo/Shutterstock.com, (right), View-point/Shutterstock.com; All other images are the property of Great Minds.

Acknowledgments

Kelly Alsup, Lisa Babcock, Cathy Caldwell, Mary Christensen-Cooper, Cheri DeBusk, Jill Diniz, Melissa Elias, Janice Fan, Scott Farrar, Krysta Gibbs, Julie Grove, Karen Hall, Eddie Hampton, Tiffany Hill, Robert Hollister, Rachel Hylton, Travis Jones, Liz Krisher, Courtney Lowe, Bobbe Maier, Ben McCarty, Maureen McNamara Jones, Cristina Metcalf, Melissa Mink, Richard Monke, Bruce Myers, Marya Myers, Geoff Patterson, Victoria Peacock, Marlene Pineda, Elizabeth Re, Meri Robie-Craven, Jade Sanders, Deborah Schluben, Colleen Sheeron-Laurie, Jessica Sims, Theresa Streeter, Mary Swanson, James Tanton, Julia Tessler, Saffron VanGalder, Jackie Wolford, Jim Wright, Jill Zintsmaster

Trevor Barnes, Brianna Bemel, Adam Cardais, Christina Cooper, Natasha Curtis, Jessica Dahl, Brandon Dawley, Delsena Draper, Sandy Engelman, Tamara Estrada, Soudea Forbes, Jen Forbus, Reba Frederics, Liz Gabbard, Diana Ghazzawi, Lisa Giddens-White, Laurie Gonsoulin, Nathan Hall, Cassie Hart, Marcela Hernandez, Rachel Hirsh, Abbi Hoerst, Libby Howard, Amy Kanjuka, Ashley Kelley, Lisa King, Sarah Kopec, Drew Krepp, Crystal Love, Maya Márquez, Siena Mazero, Cindy Medici, Patricia Mickelberry, Ivonne Mercado, Sandra Mercado, Brian Methe, Mary-Lise Nazaire, Corinne Newbegin, Max Oosterbaan, Tamara Otto, Christine Palmtag, Andy Peterson, Lizette Porras, Karen Rollhauser, Neela Roy, Gina Schenck, Amy Schoon, Aaron Shields, Leigh Sterten, Mary Sudul, Lisa Sweeney, Samuel Weyand, Dave White, Charmaine Whitman, Nicole Williams, Glenda Wisenburn-Burke, Howard Yaffe